T0213294

SpringerBriefs in Applied Sciences and Technology

SpringerBriefs present concise summaries of cutting-edge research and practical applications across a wide spectrum of fields. Featuring compact volumes of 50–125 pages, the series covers a range of content from professional to academic.

Typical publications can be:

- A timely report of state-of-the art methods
- An introduction to or a manual for the application of mathematical or computer techniques
- A bridge between new research results, as published in journal articles
- A snapshot of a hot or emerging topic
- An in-depth case study
- A presentation of core concepts that students must understand in order to make independent contributions

SpringerBriefs are characterized by fast, global electronic dissemination, standard publishing contracts, standardized manuscript preparation and formatting guidelines, and expedited production schedules.

On the one hand, **SpringerBriefs in Applied Sciences and Technology** are devoted to the publication of fundamentals and applications within the different classical engineering disciplines as well as in interdisciplinary fields that recently emerged between these areas. On the other hand, as the boundary separating fundamental research and applied technology is more and more dissolving, this series is particularly open to trans-disciplinary topics between fundamental science and engineering.

Indexed by EI-Compendex, SCOPUS and Springerlink.

More information about this series at http://www.springer.com/series/8884

João M. P. Q. Delgado · António C. Azevedo ·
Ana S. Guimarães

Drying Kinetics in Building Materials and Components

The Interface Influence

Springer

João M. P. Q. Delgado
CONSTRUCT-LFC, Department of Civil
Engineering, Faculty of Engineering
University of Porto
Porto, Portugal

António C. Azevedo
CONSTRUCT-LFC,
Faculty of Engineering
University of Porto
Porto, Portugal

Ana S. Guimarães
CONSTRUCT-LFC, Faculty of Engineering
University of Porto
Porto, Portugal

ISSN 2191-530X ISSN 2191-5318 (electronic)
SpringerBriefs in Applied Sciences and Technology
ISBN 978-3-030-31859-8 ISBN 978-3-030-31860-4 (eBook)
https://doi.org/10.1007/978-3-030-31860-4

This Springer imprint is published by the registered company Springer Nature Switzerland AG
The registered company address is: Gewerbestrasse 11, 6330 Cham, Switzerland

Preface

Moisture transfer in construction building components is fundamental for the durability of built elements, influencing the thermal behaviour and consequently the energy consumption in the building and occupants' health by the air quality. There are such of demonstrative problems due to moisture like frost/defrost damage in facades (outside), mould on interior surfaces (inside), fungus in floors and walls by rising damp (bottom), deterioration in floors and walls by food occurrences (bottom), degradation of walls and roofs by internal condensations (inside the element), etc. The solutions for treating this kind of pathologies are complex and of difficult implementation.

Considering the contact phenomena in multi-layered building components, like brick-mortar composites, or even mortar-brick-isolation-mortar solutions, the moisture transfer process and the obtained values differs from the moisture transfer considered when having different materials/layers separately. In fact, the interface promotes a hygric resistance which means that becomes a slowing of the moisture transport across the material interface. The study of liquid transport across this interface configuration (ceramic brick and mortar) implies the correct knowledge of the hygrothermal mechanical and thermal properties of the build materials employed.

This work presents an extensive experimental characterisation of two different ceramic brick blocks with a different interface, at different heights, during the drying process. First, laboratory characterisation of the building material used (ceramic bricks and different mortars) was carried out to determine their hygrothermal, mechanical and thermal properties, namely bulk porosity and density, water vapour permeability, capillary absorption, retention curve, moisture diffusivity as a function of moisture content and thermal conductivity. Finally, the moisture transfer in multi-layered systems was analysed in detail taking into account the interface contact between the building elements.

Porto, Portugal

João M. P. Q. Delgado
António C. Azevedo
Ana S. Guimarães

Contents

Chapter 1
Introduction

1.1 Motivation

Intervening in old buildings requires extensive and objective knowledge. The multifarious aspects of the work carried out on these buildings tend to encompass a growing number of specialities, with a marked emphasis on those which allow the causes of the problems that affect them to be understood, and to define appropriate treatments.

Moisture damage is one of the most important factors limiting building performance. That why the study of moisture migration in the inner parts of the materials and construction building components is of great importance for its behaviour characterization, especially for its durability, waterproofing, degradation appearance and thermal performance.

When dealing with phenomena of high complexity, it is common to consider monolithic constructive elements since the existence of joints or layers contributes to the change of moisture transfer along the respective constructive element, which will contribute to the change in mass transfer law.

In literature, although several studies exist concerning the liquid transport in multilayered porous structures, only a limited number of experimental values for the interface resistance in multilayered composites are found. The most relevant conclusions of these studies are as follows:

- The interface between layers affects the drying kinetics of the building elements;
- The transport of moisture in walls will disregard the gravitational forces since, in building materials, capillary forces are much more important due to the fine pores;
- The interface between layers delays the drying process in a much more pronounced manner, lowering the maximum flow transmitted. This flow refers to the hygric resistance of the contact between layers;

J. M. P. Q. Delgado et al., *Drying Kinetics in Building Materials and Components*, SpringerBriefs in Applied Sciences and Technology, https://doi.org/10.1007/978-3-030-31860-4_1

- Where the contact of two materials without interpenetration of the porous structure, there is no continuity of capillary pressure and interface that conditions the transport;
- The interface between layers generally causes an acceleration of the drying of the outer layer due to a lack of moisture source;
- The interface with mortar makes a hygric resistance, as well as a change in its properties. The different ways of curing of this type of interface cause different resistance values;
- When the sample has two superimposed materials of different properties, the passage of water through the interface is controlled by the rate of absorption of the second material;
- If the interface resistance is determined after capillary saturation of the first layer the change in material properties could be neglected).

1.2 Research Objectives

The objectives of this research are the study of different interfaces—hydraulic contact, perfect contact, and air space—on moisture transport during the drying process of several building materials, which were characterized previously. The importance of investigating the characteristics of these interfaces on building material is associated with the development of a new model to predict quickly and expeditiously the hygric resistance. The objectives of the study of this research are:

- Determine the material properties of ceramic red bricks, lime mortar, and cement mortar;
- Investigate the effects of different interfaces between building materials on moisture transport during the drying process;
- Quantify hygric resistances in the interface between layers, which can be experimentally determined and is extremely important to be used in advanced hygrothermal simulation programmes;
- Investigate the effects of bonding on moisture transport (hydraulic contact).

Chapter 2
Hygrothermal Properties of the Tested Materials

2.1 Introduction

This chapter describes, in detail, the hygrothermal properties of the materials tested in this work. Specific materials were introduced and then analysed, i.e., two types of ceramics and two types of mortar (cement mortar and lime mortar).

Taking into account the relevant literature on the subject, material property tests were carried out, and the obtained results were described and compared with the literature. These experiments were done to determine the physical and hygrothermal properties of the materials utilized in the study. The following properties will be presented: bulk porosity, bulk density, vapour permeability, moisture content, water absorption coefficient, moisture diffusivity, and thermal conductivity, considering the current inputs in hygrothermal models (simulations models).

In addition to the presentation of the characteristics of the materials chosen, the tests performed are also described.

2.2 Ceramic Brick

In this study, two different ceramic blocks of red brick were tested with the following size samples: ceramic A with $4 \times 4 \times 10$ cm^3 and ceramic B with $5 \times 5 \times 10$ cm^3.

2.2.1 Bulk Porosity (ε)

The porous characteristics of building materials have an important influence on the mechanisms of moisture transfer. The porous network formed by cavities and channels is the medium through which a fluid (liquid or vapour) are propagated. This

J. M. P. Q. Delgado et al., *Drying Kinetics in Building Materials and Components*, SpringerBriefs in Applied Sciences and Technology, https://doi.org/10.1007/978-3-030-31860-4_2

Fig. 2.1 Pore size
distribution

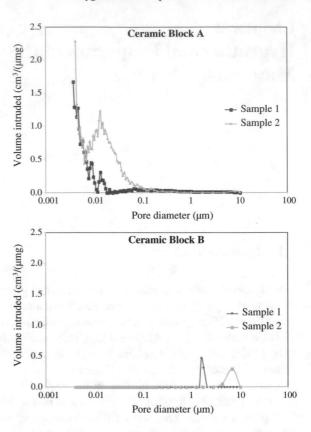

type of porosity, called open porosity, is characteristic of most building materials. In
this study, an automated mercury porosimeter (Poremaster-60 Quantachrome) and
a helium pycnometer were used. The mercury porosimeter allowed the pore size
distribution up to 0.0035 μm in diameter (see Fig. 2.1) to be determined, as well as
the total surface area normalized to the sample mass and bulk density.

Porosity, or void fraction, is a measure of the void spaces in a material and is a
fraction of the volume of voids over the total volume, between 0 and 1. The standard
ISO 10545-3 [1] for ceramic tiles was used to measure the bulk porosity of the tested
building materials (Fig. 2.2).

The samples were dried until constant mass was reached (m_1). After a period of
stabilization, the samples were kept immersed under constant pressure. The weight
of the immersed sample (m_2) and the emerged sample (m_3) the bulk porosity is given
by:

$$\varepsilon = (m_3 - m_1)/(m_3 - m_2) \tag{2.1}$$

The phenomena of transfer and fixation of humidity are strongly dependent on the
complex morpho-optic aspect of the porous space. The pores, free spaces distributed

Fig. 2.2 Material porous
structure

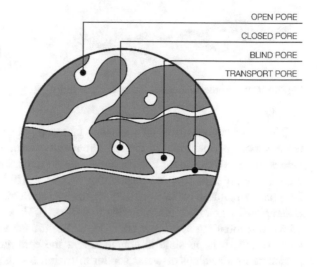

inside the solid structure, characterize the permeability of the medium, allowing the fluid flow [2]. The values obtained for the porosity of Ceramics A and Ceramics B were, respectively, 32% and 38%.

2.2.2 Bulk Density (ρ)

Several standards can be applied for the experimental determination of this property, as EN ISO 10545-3 [1] for ceramic tiles, EN 12390-7 [3] for concrete, EN 772-13 [4] for masonry units. The samples must be dried until a constant mass is reached. The samples' volume was calculated based on the average of three measurements of each dimension. The values obtained for the bulk density of Ceramics A and Ceramics B were, respectively, 1800 kg/m^3 and 1600 kg/m^3.

2.2.3 Water Vapour Permeability

The water vapour permeability, δ_p, was determined by using the cup test method, in accordance with the standard ISO 12572 [5]. The samples were cut according to the Fig. 2.3. Following the cutting process of these samples, tests were performed. The samples were sealed in a cup containing either a desiccant (dry-cup) or a saturated salt solution (wet-cup). The experiments were carried out in a climatic chamber with temperature and relative humidity controlled 23 °C and 50%. The vapour pressure gradient originates a vapour flux through the sample, and a periodic weighing of the cup allows for the calculation of the vapour transmission rate.

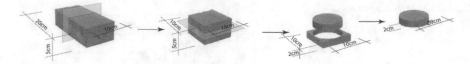

Fig. 2.3 Cutting process to water vapour permeability

The water vapour diffusion resistance factor, μ, given by $\mu = \delta_p / \delta_a$, where δ_a is the vapour permeability of the air. The results obtained showed that the average values of the water vapour diffusion resistance factor for ceramic block A samples were in the order of 33.1 (corresponding to 5.94×10^{-12} kg/msPa) for drycup and 24.9 (corresponding to 7.98×10^{-12} kg/msPa) for wetcup, while for ceramic block B samples it was 21.4 (corresponding to 9.07×10^{-12} kg/msPa) for the drycup, and 15.6 (corresponding to 12.3×10^{-12} kg/msPa) for the wet cup, respectively. From these results, it is possible to conclude that the ceramic block B samples always presented lower values of water vapour diffusion resistance when compared to the ceramic block A samples.

2.2.4 Moisture Storage Function (Moisture Content)

The moisture storage function of a specific material describes the moisture content corresponding to a specific value of relative humidity, and could be approximated by Eq. (2.2) if the values of w_{80} (moisture content at 80% relative humidity) and w_f (free moisture saturation) are known:

$$w(\phi) = w_f \frac{(b - 1) \cdot \phi}{b - \phi} \tag{2.2}$$

where ϕ is the relative humidity and b a variable.

In this study, the Gravimetric method was used to determine the moisture content at 80% relative humidity, w_{80}, in accordance with the standard ISO 12571 [6], and the free saturation moisture content was determined according to EN 12859 [7]. The samples were previously dried and the experimental values were obtained by the stabilization of specimens at different values of RH and constant temperature (23 ± 0.5 °C), by:

$$w = \frac{m_w - m_0}{m_0} \tag{2.3}$$

where m_w is the equilibrium sample mass (80% RH or saturated) and m_0 is the initial dry sample mass.

The results obtained showed that the average values of moisture content w (kg/m^3) for ceramic block A samples were in the order of 15.5 and 261.28 to $w_{80\%}$ and w_{sat}

respectively, while for ceramic block B samples it was 9.75 and 233.07 to $w_{80\%}$ and w_{sat} respectively. This shows that the ceramic block A presents the highest values of moisture content ($w_{80\%}$ and w_{sat}).

2.2.5 Water Absorption Coefficient

The experiments conducted in this test were guided by the outline of the partial immersion method described by ISO 15148 [8]. The samples were placed in distilled water with the base submerged only a few millimetres (~1–5 mm) in order to avoid build-up of hydrostatic pressure. The partial immersion method used is in accordance with the European Standard CEN/TC 89 [9]. The environmental laboratory conditions were 20 °C, 50% RH and the samples were placed in a constant-temperature water bath controlled within ±0.5 °C to avoid changes in water viscosity that might affect the absorption rate. At regular time intervals, the samples were weighed to determine the moisture uptake.

The reproducibility of the experiments was tested by independently repeating the measurement of water absorption coefficient of at least 4 identical samples (i.e., the same configuration and material), under identical operating conditions. Repeated measurements of water absorption coefficient did not differ by more than 10%.

Figure 2.4 shows the results of the average mass variation per contact area in the capillary absorption process of the samples of monolithic red brick type "A" and "B". It is possible to observe that the water absorbed curve, by capillary action, as an absorption first step (short-time faster) is directly proportional to the square root of time just as described by Bomberg [10].

The slope of this linear variation is called the water absorption coefficient (A_w) and is expressed as:

$$A_w = \frac{M_w}{A\sqrt{t}} = \frac{m_t - m_0}{A\sqrt{t}} \qquad (2.4)$$

where A_w is the water absorption coefficient (kg/m^{2}s$^{0.5}$), M_w is the total amount in time t(kg/m^2), m_t is the weight of the specimen after time (kg), m_0 is the initial mass of the specimen (kg), A is the contact area (m^2) and t is the time (s).

The absorption coefficients obtained were 0.10 kg/m^2 s$^{0.5}$ for red brick type A, and 0.19 kg/m^2 s$^{0.5}$, for red brick type "B", respectively. The absorption coefficient for red brick type "B" is approximately two times higher, an expectable result as the density of red brick type "B" is greater than the density of red brick type "A". These results are in accordance with other experimental values found in literature for the same building materials, with a range of experimental values between 0.05 and 0.29 kg/m^{2}s$^{0.5}$ (see [11–16]).

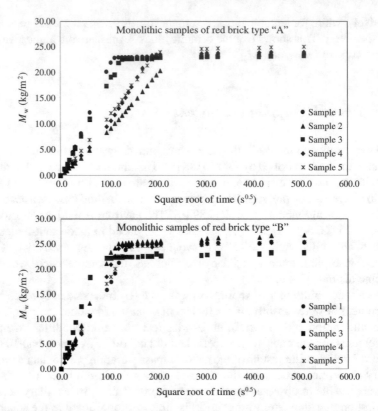

Fig. 2.4 Capillary absorption curve of monolithic samples of red brick type "A" and "B", partially immersed in water, as a function of the square root of time

2.2.6   *Moisture Diffusivity*

Moisture diffusivity is a transport property that is frequently used in the hygrothermal analysis of building envelope components. This property is dependent on the local moisture content. The experiments that lead to the detailed information on the dependence of diffusivity on moisture content are often very sophisticated.

The experimental wetting and drying data to determine the diffusivity was interpreted by using Fick's diffusion model. In a one-dimensional formulation, with the diffusing substance moving in the normal direction to a sheet of medium of thickness *L,* the diffusion equation can be written as [17]:

$$\frac{\partial w}{\partial t} = \frac{\partial}{\partial x}\left(D_{\mathrm{w}}(w)\frac{\partial w}{\partial x} \right) \tag{2.5}$$

subject to the following boundary conditions,

$$t = 0 \quad 0 < x < \infty \quad w = w_0 \tag{2.6a}$$

$$t > 0 \quad x = 0 \quad w = w_\infty \tag{2.6b}$$

$$t > 0 \quad x \to \infty \quad w = w_0 \tag{2.6c}$$

The analytical solution of Eq. (2.5) with the initial and boundary conditions (2.6a)–(2.6c), if constant diffusion coefficient is assumed ($D_w(w) = D_{\text{eff}}$), is:

$$\frac{w - w_0}{w_\infty - w_0} = 1 - erf\left(\frac{x}{2\sqrt{D_{\text{eff}}t}}\right) \tag{2.7}$$

Integrating in respect of t the rate of the penetration of sample face unit area ($x = 0$) by water vapour, the total amount of diffusing substance in time t is obtained, then the appropriate solution of the diffusion Eq. (2.5) may be written as follows:

$$MR = \frac{8}{\pi^2} \times \sum_{n=0}^{\infty} \frac{1}{(2n+1)^2} \exp\left[-(2n+1)^2\pi^2\frac{D_{\text{eff}}t}{L^2}\right] \tag{2.8}$$

and long drying times (neglecting the higher order term by setting $n = 0$) has been simplified as follows:

$$\ln(MR) = \ln\left(\frac{8}{\pi^2}\right) - \left[\pi^2\frac{D_{\text{eff}}t}{L^2}\right] \tag{2.9}$$

Finally, the effect of temperature on diffusivity is described using Arrhenius relationship:

$$D_{\text{eff}} = A_0 \times \exp\left(-\frac{E_a}{RT}\right) \tag{2.10}$$

However, the variations of $D_w(w)$ with the moisture content w are usually nonlinear for most common porous media and there are no analytical solutions. For one-dimensional water absorption with $w = w_0$ at $x > 0$ and $t = 0$, and $w = w_\infty$ at $x = 0$. and $t > 0$. Equation (2.5) can be expressed in terms of a single variable, proposed by Boltzmann, $\eta = x/\sqrt{t}$ [18]:

$$-\frac{\eta}{2}\frac{dw}{d\eta} = \frac{d}{d\eta}\left(D_w(w)\frac{dw}{d\eta}\right) \tag{2.11}$$

with the following boundary conditions

$$\eta = 0 \quad w = w_\infty \tag{2.12a}$$

$$\eta \to \infty \quad w = w_0 \tag{2.12b}$$

and $D_w(w)$ can be determined by integrating Eq. (2.11) using the boundary conditions (2.12a)–(2.12b).

$$D_{\mathrm{w}}(w) = -\frac{1}{2}\frac{d\eta}{dw}\int_{w0}^{w} \eta dw \tag{2.13}$$

In this study, the identification of moisture diffusivity of different ceramic bricks, at constant temperature and as a function of moisture content, is presented. For the investigation of moisture diffusivity, the lateral sides of the samples were insulated by water-proof and vapour-proof epoxy resin. The front side of the specimens was placed in contact with water. Afterword's, the moisture profiles were measured in the climatic chamber undersothermal conditions (25 °C) in order to eliminate secondary effects on moisture transport. After the experiment, the moisture profiles of each sample were measured using the gamma ray method.

Figure 2.5 shows an example of the moisture content versus Boltzmann variable, η. For the tested materials, the Boltzmann transformation seems to hold very well. Based on this kind of η profiles the moisture diffusivity can be determined.

The moisture diffusivity values presented in Fig. 2.5 were obtained by the Boltzmann transformation method over the whole moisture content. It is possible to observe that the shape of the curve representing the dependence of diffusion coefficients with moisture content was similar in all materials tested, and in accordance with literature results [18–21]. The moisture diffusivity first decreases with increasing moisture content in the hygroscopic region, passes a minimum and then increases, with an almost exponential behaviour, in the capillary range at high moisture content.

2.2.7   Thermal Properties

Thermal conductivity could be measured by different methods that are broadly classified as steady-state or transient methods. In this study, the measurements were performed using a transient method, the thermal shock probe method (CT-Mètre), in accordance with the standard protocol defined in ISO 8301 [22]. The thermal conductivity values obtained for the samples tested, ceramic blocks A and B, were presented in Fig. 2.6, in function of moisture content.

The values obtained for the 2 ceramic blocks tested are very similar and in accordance with the values reported by Taoukil et al. [23] and Koci et al. [21] who presents values of ceramic blocks in the range between 0.40 and 0.70 W/mK, for different values of moisture content.

The properties of the ceramic blocks tested were measured carefully and the obtained values are summarized, in Table 2.1.

Fig. 2.5 a Example of
moisture profiles plotted as a
function of $\eta = x/t^{0.5}$;
b moisture diffusivity
coefficient as a function of
the moisture content

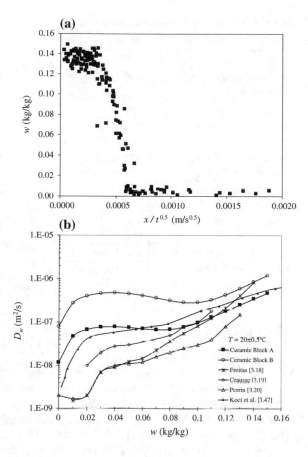

(a)

(b)

$T = 20\pm0.5°C$
- Ceramic Block A
- Ceramic Block B
- Freitas [3.18]
- Crausse [3.19]
- Perrin [3.20]
- Kočí et al. [3.47]

Fig. 2.6 Thermal
conductivity in function of
moisture content

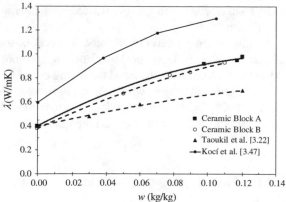

- Ceramic Block A
- Ceramic Block B
- Taoukil et al. [3.22]
- Kočí et al. [3.47]

Table 2.1 Properties of the ceramic blocks tested

Properties	Block A	Block B
Apparent density, ρ, (kg/m^3)	1800	1600
Water absorption coefficient, A (kg/m^2 s$^{0.5}$)	0.10	0.19
Porosity, ε, ($-$)	0.32	0.38
Water vapor permeability, δ_p (kg/msPa)	5.94×10^{-12} (dry) 7.98×10^{-12} (wet)	9.07×10^{-12} (dry) 12.3×10^{-12} (wet)
Water vapor diffusion resistance factor, μ ($-$)	33.1 (dry cup) 24.9 (wet cup)	21.4 (dry cup) 15.6 (wet cup)
Diffusion-equivalent air-layer thickness, s_d (m)	0.61 (dry cup) 0.48 (wet cup)	0.37 (dry cup) 0.29 (wet cup)
Moisture content, w (kg/m^3)	$w_{80\%} = 15.51$ $w_{sat.} = 261.38$	$w_{80\%} = 9.75$ $w_{sat.} = 233.07$
Thermal conductivity curve, λ (W/mK), w (kg/kg))	$\lambda = 0.40$ ($w = 0.00$)	$\lambda = 0.38$ ($w = 0.00$)
	$\lambda = 0.91$ ($w = 0.10$)	$\lambda = 0.68$ ($w = 0.05$)
	$\lambda = 0.96$ ($w = 0.12$)	$\lambda = 0.92$ ($w = 0.11$)

2.3 Mortars

Masonry mortar is used for wall construction. The dosage used in mortar is one of the most important factors for the workability associated with the rheological aspects of mortars. For the dosage used in this research, two references were consulted: NBR 7200 [24] and ASTM C 270 [25]. The two standards have in common the same proportion of sand binder, i.e., 1:3 (in volume). Thus, in mortar contact specimens, two types of mortar were selected: lime mortar with a proportion of 1:3 (lime, sand) and cement mortar with a ratio of 1:1:6 (cement, lime and sand). The settling thickness was 10 mm.

Mortar placement between the blocks was done through two wooden jigs on the side faces of the blocks, secured by a hook that promotes the contraction of the wood to keep them fixed and thus avoid the laying of the upper block on the mortar.

As an attempt to reproduce mortars used in masonry walls and for supposed reproducibility in other investigations, the experimental characterization of mortars tested (cement and lime mortar) and the mixture aggregates are described below.

2.3.1 Fine Aggregate Analyse

The characterization of the fine aggregates (sand) used in the preparation of mortars took into account the following tests and their respective Brazilian standards:

- Granulometry of the fine aggregate—NBR NM 248 [26]
- Dry aggregate specific gravity (Flask of Chapman)—NBR 52 [27]
- Aggregate specific gravity in (SSS)[3] conditions—NBR 52 [27]
- Bulk density—NBR 52 [27]
- Determination of the fine materials content—NBR NM 46 [28]
- Fineness modulus—NBR NM 248 [26]
- Maximum diameter—NBR NM 248 [26]
- Determination of the unit mass—NBR NM 45 [29]
- Water absorption—NBR NM 30 [30].

The sand used was acquired in the Metropolitan Region of Recife and all the tests were carried out in the Laboratory of Construction Materials of Catholic University of Pernambuco—TECOMAT, Brazil.

2.3.1.1 Granulometric Composition

The determination of the granulometric composition aims to classify the aggregate as a function of the size of the grains. The test method described in NBR NM 248 [26] is to determine if it is necessary to collect two samples of the aggregate to be analysed which must then be washed and preheated to a temperature of $(105 + 5)$ °C. Classification occurs through a set of sieves with openings standardized by ABNT (see Fig. 2.7). The device is assembled in a decreasing manner so that the sieves of larger apertures overlap the sieves of smaller apertures; the material is then sieved so that each fraction is retained in the sieves, and then separated and weighed. Figure 2.7 also illustrates the tools used in the experiment.

The results obtained are presented in the following tables and figures (see Tables 2.2 and 2.3 and Figs. 2.8 and 2.9).

Figures 2.8 and 2.9 present a comparison between the obtained results for sands utilized in the mortar mixtures for the optimal zone, usable and not usable, or not advisable. The desirable granulometric standard is that found in the optimal region.

Fig. 2.7 Sieves used

Table 2.2 Granulometry of coarse sand aggregate—NBR NM 248 [26]

Opening of the sieves (mm)	Mass retained (g)		Average of retained mass (g)	Average of retained mass (%)	Cumulated of retained mass (%)
	Test n° 1	Test n° 2			
9.5	6.0		6.08	0.8	1
6.3	11.1	11.06	11.08	1.5	2
4.75	13.3	13.34	13.32	1.8	4
2.36	74.0	74.11	74.06	10.1	14
1.18	196.2	196.14	19.17	26.7	41
0.6	284.4	284.43	284.42	38.7	80
0.3	123.3	123.26	123.28	16.8	96
0.15	19.6	19.64	19.62	2.7	99
Bottom	7.5	7.43	7.47		
Total	735.4	735.6	735.5		

Table 2.3 Granulometry of fine sand aggregate—NBR NM 248 [26]

Opening of the sieves (mm)	Mass retained (g)		Average of retained mass (g)	Average of retained mass (%)	Cumulated of retained mass (%)
	Test n° 1	Test n° 2			
25	0	0	0	0	0
19	0	0	0	0	0
12.5	0	0	0	0	0
9.5	0	0	0	0	0
6.3	0	0	0	0	0
4.75	1.0	1.08	1.04	0.1	0
2.36	2.1	2.16	2.13	0.3	0
1.18	9.7	9.62	9.66	1.3	2
0.6	274.4	274.33	274.37	37.2	39
0.3	249.0	249.09	249.05	33.8	73
0.15	112.6	212.55	112.58	15.3	88
Bottom	88.8	88.87	88.84		
Total	737.6	737.7	737.7		

In other words, it presents a proportional standard with equality between the grains. If this standard presents different proportional grades, this will result in a granulometric standard which falls in the usable and not usable regions, depending on the differentiation grade of the grains. Knowing that, the granulometric curves of the employed sands for the mortar formulation lead us to conclude that two different curve patterns are needed. Thus, when combining both, it is possible to find the

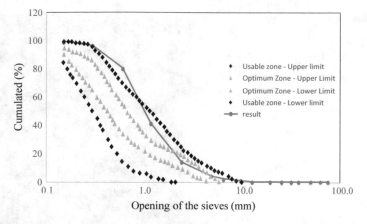

Fig. 2.8 Granulometry curve of coarse sand aggregate

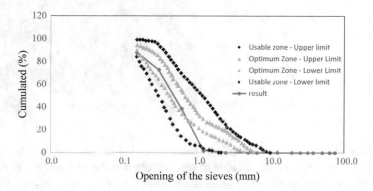

Fig. 2.9 Granulometry curve of fine sand aggregate

predominant granulometry in the optimal region, which allows us to obtain a better quality of mortar.

2.3.1.2 Specific Mass

In order to determine the specific mass of the fine aggregate, the procedure described in the Brazilian standard NBR NM 52 [27], using the Chapman vial (see Fig. 2.10), a standardized vial that indicates the displacement of the water column volume after the materials insertion, was used. The material specific mass is the ratio of its mass and the volume occupied, excluding voids between the grains.

The sand was oven dried at a temperature of approximately 105 °C and a sample of 500 g was withdrawn. This sample was then inserted into the Chapman vessel, which was already filled with water to a volume of 200 cm^3; as the material was

Fig. 2.10 Chapman vessel

being placed in the vessel, the water column ascended so that, the variation of this displacement represented the volume of sand inserted at the end of the process.

As expected, the results showed an increase of the drying time constant for the materials with perfect contact interface, compared to the monolithic materials, and that the further away from the base the interface is located, the greater the drying time constant. Summarizing, the interface could significantly retard the flow of moisture transport. For example, the discontinuity of moisture content across the interface indicated that there was a difference in capillary pressure across the interface.

$$\gamma = \frac{m}{(L - 200)} \qquad (2.14)$$

In Eq. (2.14), m is the dry mass of the material and L is the final volume of the water column, the result is expressed in g/cm^3.

2.3.1.3   Fine Materials Content

Fine materials present in the aggregate, i.e., those passing through the 75 μm aperture sieve, also called powder material, should be analysed by the method described in NBR NM 46 [28]. Its quantity, when higher than the one foreseen in the standard

NBR 7211 [31] that is of 5%, can harm the mixture, either of concrete or mortar, because the very fine grains make difficult the adhesion of the cement paste to the aggregate.

The sample was oven-dried at 110 °C until mass constancy, after which a 100 g sample was withdrawn. This quantity was placed on the sieves with opening 1.2 mm and 75 μm and subjected to the washing in successive times to eliminate the fine material adhered to the aggregate with the water. The process was completed when the water passed through was completely clean. The material was again placed in an oven to evaporate the water and obtain the final mass. The result was calculated according to the expression:

$$Powder\ material\ content = \frac{M_i - M_f}{M_i} \times 100 \qquad (2.15)$$

where M_i is the initial mass and M_f is the final mass.

2.3.1.4 Unit Mass Test

The unit mass in the loose state of the fine aggregate is determined by NBR NM 45 [29]. This method requires a cylindrical vessel with a known volume of 20 l, 16 mm diameter metal bar, a concrete shovel, metal ruler, a balance of 50 g and a quantity of dry sand sufficient to occupy the volume of the vessel.

In this test, method C was used. The dry sand is added without compaction until the entire volume of the vessel is occupied, a metal ruler is used to remove the excess sand in the vessel and the sand set plus the vessel are weighed and the result is calculated according to the expression:

$$P_{ap} = \frac{(m_{ar} - m_r)}{V} \qquad (2.16)$$

where ρ_{ap} is the unit mass of the aggregate (kg/m^3), m_{ar} is the mass of sample plus container (kg), m_r is the empty container mass (kg) and V is the container volume (m^3).

2.3.1.5 Water Absorption

This test aims at determining the level of water absorption in small aggregates for specific use. It's extremely relevant to obtain specific information concerning the degree of absorption from the aggregate which will be used in a concrete trace. In this case, when the aggregate already shows some percentage of moisture, the amount of water usually added to the paste should be decreased. Therefore, an eventual segregation will be prevented, where larger particles will separate from smaller ones, thus causing the emergence of layers in the concrete and consequently producing a

Table 2.4 Experimental results of the aggregate characterization

	Dry aggregate specific gravity (g/cm^3)	Aggregate specific gravity in (SSS)^3 conditions (g/cm^3)	Bulk density (g/cm^3)	Powder material (%)	Fineness modulus	Maximum diameter (mm)	Water absorption (%)
Fine aggregate	2.53	2.55	2.60	0.5	3.35	4.75	1.2
Coarse aggregate	2.42	2.44	2.48	2.7	2.02	2.02	1.0
	NBR NM 52 [27]	NBR NM 52 [27]	NBR NM 52 [27]	NBR NM 46 [52]	NBR NM 248 [26]	NBR NM 248 [26]	NBR NM 30 [30]

serious reduction in resistance. The test method, described in NBR NM 30 [30], consists of quartering 1 kg of an aggregate sample which must be washed and placed in advance in the greenhouse at a temperature of (105 ± 5) °C for 6 h.

Tables 2.2 and Table 2.3 present the experimental results of the granulometric composition obtained and the experimental results of the aggregate characterization are described in Table 2.4.

The results showed that the powder materials present values of less than 5% for fine and coarse sand aggregates. The granulometric distribution of the two types of sand analysed is a guarantee of good sand compaction.

2.3.2   Cement Portland

The cement mortar used in making samples was of the type CPII F 32 which was provided in 50 kg sacks. It was purchased in a warehouse in the Metropolitan Region of Recife, Brazil. The reference values for this product, provided by the cement factory CIMPOR, are given in Table 2.5.

2.3.3   Hydrated Lime

The lime used to make the mortar (laying and coating) was the of type CHII of Dolomil, supplied in 20 kg sacks. Its features can be observed in Table 2.6, which were given be the quality control department of Dolomil industry.

Table 2.5 Cement Portland characterization

Test	Method	Result	Specification
Fineness #75 μm (%)	NBR 11579 [32]	2.1	$\leq$12
Specific surface (cm^2/g)	NBR NM 16372 [33]	4955	$\geq$2600
Start of curing time (min)	NBR NM 65 [34]	182	$\geq$60
End of curing time (min)	NBR NM 65 [34]	254	$\leq$600
Expandability (mm)	NBR 11582 [35]	0	$\leq$5
Compressive strength (MPa)	NBR 7215 [36]	21.1 (3 days)	$\geq$10.0
		26.3 (7 days)	$\geq$20.0
		34.7 (28 days)	$\geq$32.0
Magnesium oxide—MgO (%)	FRXPA	5.7	$\leq$6.5
Sulphuric anhydride—SO$_3$ (%)	FRXPA	2.84	$\leq$4.0
Calcium oxide—CaO (free) (%)	FRXPA	1.40	X

Table 2.6 Hydrated lime characterization

Test		Result	Specification
Relative Humidity (110 °C) (%)		0.75	X
Loss on ignition (1000 °C) (%)		27.35	X
Carbon dioxide (%)		3.60	$\leq$7.0
Sulphuric anhydride (%)		0.03	X
Total calcium oxide (%)		42.38	X
Magnesium oxide (%)		27.23	X
CaO + MgO not hydrated (%)		2.71	$\leq$15.0
Total oxides na base de non-volatile (%)		95.82	$\geq$88.0
Water retention (%)		97.64	$\geq$75.0
Particle size (%)	(+) 0.600 mm	0.0	$\leq$0.5
	(+) 0.075 mm	11.85	<15.0

2.3.4 Mortar Properties

2.3.4.1 Mixture

The mortars were submitted to an analysis process in which their properties were evaluated both in the fresh state and in the hardened state by diversifying traces with and without the cement addition. The mixing procedure had been previously determined and consisted of manual mixing of the dried binder and aggregate. Next, water was added while blending the mixture for 5 min in a low-speed automatic mixer. The amount of water used in the mixture was pre-determined with the aim of achieving the same consistency in both mixes (reference cement mortar and lime

Table 2.7 Mortars mixtures analysed

Type	Proportion	Mixture				
		Cement	Lime	Fine sand	Coarse sand	Water
Cement mortar	Volume	1	1	3	3	
	Mass (g)	228	198	1058	1133	460
Lime mortar	Volume	0	1	1.5	1.5	
	Mass (g)	0	396	1058	1133	415

mortar). The water binder proportion is 1.04, the same used in Nunes and Slížková [37]. Table 2.7 shows the types of mortars studied.

2.3.4.2   Physical Properties

The fresh-state mortar was featured through tests showing consistency, density and to the content of indoor air. The mortar consistency can be evaluated through the procedures described in standard NBR 13276 [38], which allow verification of the degree of plasticity being supported by a consistency table ("Flow Table") as illustrated in Fig. 2.11.

Samples are exposed to successive falls of a predetermined height. Thus, the more plastic the mortar, the bigger its final diameter will be.

Mortar can be considered dry when the consistency index (flow table) is below 250 mm (e.g., mortar for sub-floor). When the consistency index of mortar is between 260 and 300 mm (e.g., plaster mortar), the mortar is considered to be plastic. Finally, mortar whose consistency index is above 360 mm is considered to be of a very plastic nature, (for example, roughcast mortar). The achieved results are summarized in Table 2.8.

Fig. 2.11 Consistency evaluation

Table 2.8 Values of the consistency index

Mortar	Mixture	Test 1	Test 2	Average (mm)
Cement mortar	1:1:6	270.2	271.3	270.75
Lime mortar	0:1:3	178.8	178.3	178.55

Table 2.9 Experimental
values of density of the fresh
mortar and air content
incorporated

Mortar	Mixture	Fresh mortar density (g/cm^3)	Air content incorporated $(\%)$
Cement mortar	1:1:6	1.95	35
Lime mortar	0:1:3	2.07	29

The mortar density in the fresh state and the content of indoor air, obtained in accordance with the procedures described in NBR 13278 [39], are presented in Table 2.9.

2.3.4.3 Mechanical Properties

The mortars in the hardened state were characterized by the following tests:

- Axial compressive strength (rupture test), NBR 13279 [40]
- Tensile strength in flexion, NBR 13279 [40].

In order to measure the axial compressive strength, three cylindrical specimens were prepared for each mortar type, and to measure the tensile strength in flexion three more were used for each type with dimensions of $4 \times 4 \times 16\,cm^3$ (see Fig. 2.12). The specimens were cured for a period of 28 days, before carrying out the resistance tests, as showed in Fig. 2.13 and described in Tables 2.10 and 2.11.

Tables 2.10 and 2.11 show the experimental results obtained for axial compression strength and tensile strength of two different mortar mixture, 1:1:6 and 0:1:3, respectively. The results showed that the cement mortar analysed is of type "N" (see Table 2.12), in accordance with the mixture and resistance (see ASTM C270 [25]).

Related to the mortar without cement Portland (see Table 2.13), comparing to Fort et al. [41] gave 0.91 to compressive strength and 0.4 strength in flexion; Vejmelková et al. [42] as opposed to 1.5 to compressive strength and 0.7 strength in flexion.

Fig. 2.12 Mortar moulding and demoulding

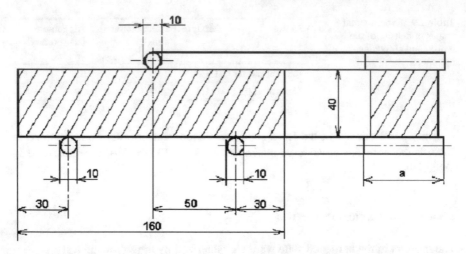

Fig. 2.13 Loading device for determination of tensile strength in bending

Table 2.10 Axial compression strength and tensile strength, for a mortar mixture of 1:1:6

Mortar	Sample	Axial compressive strength (kN)	Tensile strength (MPa)	Tensile strength in flexion (MPa)
Mixture 1:1:6	1	164.8	10.3	2.6
	2	160.0	10.0	3.2
	3	153.6	9.6	3.0
	Average		10.0	2.9
	Standard deviation		0.4	0.3
	Variation coefficient (%)		4	10

Table 2.11 Axial compression strength and tensile strength, for a mortar mixture 0:1:3

Mortar	Sample	Axial compressive strength (kN)	Tensile strength (MPa)	Tensile strength in flexion (MPa)
Mixture 0:1:3	1	20.8	1.3	0.55
	2	20.8	1.3	0.57
	3	20.8	1.3	0.55
	Average		1.3	0.60
	Standard deviation		0	0
	Variation coefficient (%)		0	2

Table 2.12 Properties of the mortars tested

Properties	Cement mortar	Lime mortar
Apparent density, ρ, (kg/m^3)	1878	1810
Water absorption coefficient (kg/m^{2}s$^{0.5}$)	0.15	0.12
Porosity, ε, (−)	0.20	0.21
Water vapor permeability, δ_p (kg/msPa)	0.81×10^{-11} (dry) 1.09×10^{-11} (wet)	1.35×10^{-11} (dry) 1.50×10^{-11} (wet)
Water vapor diffusion resistance factor, μ (−)	23.9 (dry cup) 17.9 (wet cup)	14.25 (dry cup) 12.8 (wet cup)
Diffusion-equivalent air-layer thickness, s_d (m)	0.50 (dry cup) 0.38 (wet cup)	0.30 (dry cup) 0.26 (wet cup)
Moisture content, w (kg/m^3)	$w_{80\%} = 38.16$ $w_{sat.} = 228.59$	$w_{80\%} = 8.16$ $w_{sat.} = 186.61$
Thermal conductivity, λ (W/mK)	0.786	0.799

2.3.4.4 Hygrothermal Properties

The bulk porosity, bulk density, water vapour permeability, moisture storage function (moisture content), water absorption coefficient, moisture diffusivity and thermal property were determined in accordance to the standards described for the monolithic samples of ceramic blocks.

It is important to notice that there is a relationship between the mortars and the sample properties in masonry performance related to water penetration resistance. The mortar properties considered in the study include compressive strength, workability, water retention and porosity. The properties of masonry elements include surface texture, porosity and initial rate of absorption [43].

Since the building walls present mortars with average values of porosity and large amount of pores that allow the capillary action and movement of water inside the coating system, the presence of moisture can be observed, mainly in the lower part of the walls. Mortars are characterized by the presence of voids within the materials (pores), which may vary according to the water/cement ratio, as well as the granulometry of the aggregates used in the mortar [44].

Bulk Porosity

Figure 2.14 presents the results of the porosimetry analysis of the mortars tested and it is possible to observe the differences between the 2 mortars analysed, even taking into account that the materials present similar bulk porosity.

Finally, Fig. 2.14 shows the pore size distributions of cement mortar and lime mortar samples. Both distributions reach the minimum measurable diameter value

Table 2.13 Properties of the blocks (A and B) of the mortars tested with the literature results

	Material	Proportion (C:L:S:W/C)	Consistence	Apparent density (kg/m³)	Compressive strength (Mpa)	Flexural strength (Mpa)	Thermal Conductivity Test (W/m.k)	Porosity (%)	Wsat (Kg/m³)	A (kg.m^{-1}.s$^{-1.2}$)	Water vapor permeability (dry cup) (kg/msPa)	Water vapor diffusion resestance factor (dry cup)	Water vapor diffusion resestance factor (wet cup)
Block A	**Brick**	NA	NA	**1800.00**	**ND**	**ND**	**0.40–1**	**0.32**	**261.38**	**0.1**	**5.94E−12**	**33.1**	**21.4**
Block B	**Brick**	NA	NA	**1600.00**	**ND**	**ND**	**0.39–0.97**	**0.38**	**233.07**	**0.19**	**9.07E−12**	**24.9**	**15.6**
Roels et al. [13]	Brick	NA	NA	2002.00	ND	ND	ND	23.81	147.9	0.16	ND	ND	ND
García et al. [14]	Brick	NA	NA	ND	ND	ND	ND	ND	ND	0.155	ND	ND	ND
García et al. [14]	Brick	NA	NA	ND	ND	ND	ND	ND	ND	0.294	ND	ND	ND
Freitas [18]	Brick	NA	NA	1926.00	ND	ND	1.00	28.00	78.9	ND	ND	ND	ND
Fernandes and Lourenço [47]	Brick	NA	NA	1742.00	ND	ND	ND	33.00	ND	ND	ND	ND	ND
Fernandes and Lourenço [47]	Brick	NA	NA	1754.00	ND	ND	ND	26.3	ND	ND	ND	ND	ND
Fernandes and Lourenço [47]	Brick	NA	NA	1800.00	ND	ND	ND	28.2	ND	ND	ND	ND	ND
Fernandes and Lourenço [47]	Brick	NA	NA	1747.00	ND	ND	ND	29.2	ND	ND	ND	ND	ND

(continued)

Table 2.13 (continued)

Material	Proportion (C:L:S:W/C)	Consistence	Apparent density (kg/m³)	Compressive strength (Mpa)	Flexural strength (Mpa)	Thermal Conductivity Test (W/m.k)	Porosity (%)	Wsat (Kg/m³)	A (kg.m⁻¹.s⁻¹.²)	Water vapor permeability (dry cup) (kg/msPa)	Water vapor diffusion resestance factor (dry cup)	Water vapor diffUsion resestance factor (wet cup)
Ferrandes and Lourenço [47]	NA	NA	1739.00	ND	NE	ND	30.4	ND	NE	ND	ND	ND
Fernandes and Lourenço [47]	NA	NA	1656.00	ND	ND	ND	27.5	ND	ND	ND	ND	ND
Václav Kočí et al. [21]	NA	NA	1831.00	ND	ND	0.59–1.73	27.9	ND	0.26	ND	22.1	8.8
Nunes et al. [48]	NA	NA	ND	ND	ND	ND	33.00	ND	ND	ND	ND	ND
Künzel [49]	NA	NA	1600.00	ND	ND	ND	ND	ND	ND	ND	9.5	8.00
Derluyn et al. [30]	NA	NA	2087.00	ND	ND	ND	21.00	130.00	0.116	6.09E–12	ND	ND
Janssen et al. [:1]	NA	NA	ND	ND	ND	ND	ND	270.00	0.43	ND	ND	ND
N. Laaroussi et al. [51]	NA	NA	1777.00	ND	ND	0.35	ND	ND	ND	ND	ND	ND
Laaroussi et al. [51]	NA	NA	1652.00	ND	ND	0.35	ND	ND	ND	ND	ND	ND
Suchorab et al. [52]	NA	NA	2450.00	ND	ND	0.81	23.7	ND	0.07	ND	10.3	ND
Suchorab et al. [52]	NA	NA	1590.00	ND	ND	ND	34.5	ND	ND	ND	ND	ND

(continued)

Table 2.13 (continued)

Material	Proportion (C:L:S:W/C)	Consistence	Apparent density (kg/m³)	Compressive strength (Mpa)	Flexural strength (Mpa)	Thermal Conductivity Test (W/m.k)	Porosity (%)	Wsat (Kg/m³)	A (kg.m^{-1}.s$^{-1,2}$)	Water vapor permeability (dry cup) (kg/msPa)	Water vapor diffusion resestance factor (dry cup)	Water vapor diffUsion resestance factor (wet cup)
Rêgo [53]	Brick	NA	NA	2100.00	ND	ND	ND	ND	ND	0.0686	ND	ND	ND
Azevedo [15]	Brick	NA	NA	ND	ND	ND	ND	ND	ND	0.145	ND	ND	ND
Gonçalves et al. [54]	Brick	NA	NA	ND	ND	ND	ND	0.24	ND	0.13	ND	ND	ND
Depraetere et al. [55]	Brick	NA	NA	2005.00	ND	ND	ND	0.24	240.00	0.184	ND	ND	ND
Cement mortar 1	**1:6:1.12 (p)**	**271**	**1878.00**	**10.00**	**2.9**	**0.79**	**20.00**	**228.59**	**0.145**	**8.10E−12**	**23.90**	**17.90**
Corinaldesi et al. [56]	Cement mortar 1	0.2:2.8:0.6 (m)	630	2191.00	32.00	12.00	0.73	ND	ND	ND	ND	ND	ND
Carasek et al. [57]	Cement mortar 1	2:9:ND (p)	ND	ND	ND	ND	ND	ND	ND	ND	ND	ND	ND
Demirboğa and Gül [58]	Cement mortar 1	0:2:0.5 (m)	ND	2040.00	ND	ND	1.18	ND	ND	ND	ND	ND	ND
Xu and Chung [59]	Cement mortar 1	0:1:0.5 (na)	ND	2040.00	ND	ND	0.58	ND	ND	ND	ND	ND	ND
Fort [41]	Cement mortar 1	0:3:0.5 (m)	ND	1755.00	0.91	0.4	ND	33.00	ND	0.175	1.63E−11	10.2	ND
Künzel [49]	Cement mortar ND		ND	1900.00	ND	ND	ND	25.00	ND	ND	ND	19.00	18.00
Wilson et al. [60]	Cement mortar 1:0:2:0 (na)		ND	1 ND	ND	ND I	ND	ND	ND	ND	ND	ND	ND

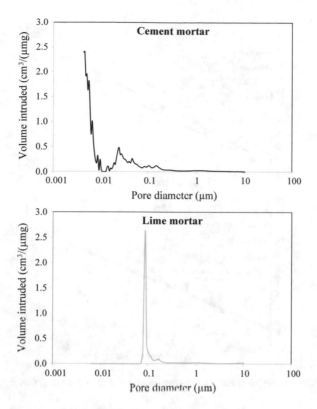

Fig. 2.14 Porosimetry results of the mortars tested

of 3.5 nm, which means that the equipment is not capable of measuring all mesopores and micropores without interfering with the main conclusions in the pore range studied. However, it is possible to calculate from the total percentage of pores resorting to the apparent and true densities. The values obtained for the porosity of Cement mortar and Lime mortar were, respectively, 20% and 21%.

Water Absorption

The experiments conducted in this test were guided by the outline of the partial immersion method described by ISO 15148 [8]. The samples were placed in distilled water, with the base submerged only a few millimetres (~1–5 mm) in order to avoid build-up of hydrostatic pressure. The environmental laboratory conditions were 20 °C and 40% RH and the samples were placed in constant-temperature water bath controlled within ±0.5 °C to avoid changes in water viscosity that might affect the absorption rate. At regular time intervals, the samples were weighed to determine the moisture uptake.

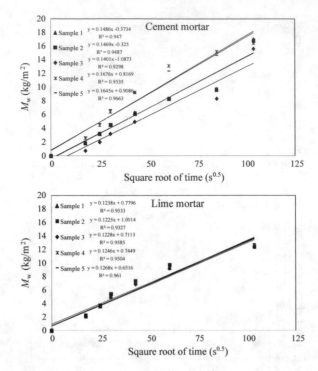

Fig. 2.15 Capillary absorption curve of cement mortar and lime mortar in function of the square root of time

The reproducibility of the experiments was tested by independently repeating the measurement of water absorption coefficient of at least 6 identical samples (i.e. the same configuration and material), under identical operating conditions, and repeated measurements of water absorption coefficient did not differ by more than 10%.

Figure 2.15 shows the results of the average mass variation per contact area in the capillary absorption process for the mortar samples tested. It is possible to observe that the water absorbed curve by capillary action, as an absorption first step (short-time faster), is directly proportional to the square root of time just as described by Eq. (2.4).

The experimental assessment of the water absorption coefficients showed a clear difference (twice) of values between the two mortars tested. The averages values obtained are 0.145 ± 0.09 kg/(m^2 $s^{0.5}$) and 0.124 ± 0.01 kg/(m^2 $s^{0.5}$), for cement mortar and lime mortar, respectively. The coefficient of variation found for each set of identical samples was approximately 15%. In comparing, for lime mortar 0.17 [41] and 0.20 [42].

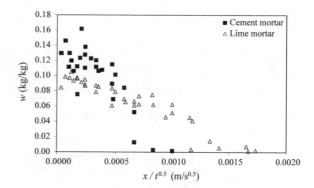

Fig. 2.16 Example of moisture profiles plotted as a function of $\eta = x/t^{0.5}$

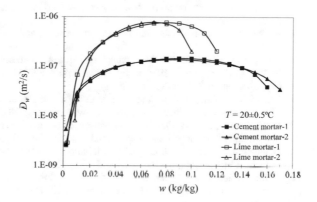

Fig. 2.17 Moisture diffusivity coefficient as a function of the moisture content

Moisture Diffusivity

Figure 2.16 shows the moisture content versus Boltzmann variable, η, for cement and lime mortar. For the materials tested, the Boltzmann transformation seems to hold very well. Based on this kind of η profiles the moisture diffusivity can be determined.

The moisture diffusivity values presented in Fig. 2.17 were obtained by the Boltzmann transformation method over the whole moisture content. It is possible to observe that the shape of the curve representing the dependence of diffusion coefficients with moisture content was similar in all materials tested, and in accordance with literature results of Perrin [20], Daian [45] and Fernandes and Philippi [46].

Table 2.12 describes the experimental results obtained for cement mortar and lime mortar. The results show that the cement mortar presents the highest values of moisture content ($w_{80\%}$ and w_{sat}), bulk density and lower values of water vapour permeability.

The density, porosity, thermal conductivity, water vapour permeability, and water vapour diffusion resistance factor are practically identical and in accordance with the results found in literature, as Table 2.13.

References

1. ISO 10545-3. Ceramic tiles—Part 3: Determination of water absorption, apparent porosity, apparent relative density and bulk density, Geneva, Switzerland (1995)
2. N. Mendes, P.C. Philippi, A method for predicting heat and moisture transfer through multilayer walls based on temperature and moisture content gradients. Int. J. Heat Mass Transf. **48**, 37–51 (2005)
3. EN 12390-7. Testing hardened concrete. Density of hardened concrete, Geneva, Switzerland (2009)
4. EN 772-13. Methods of test for masonry units. Determination of net and gross dry density of masonry units, except for natural stone, Geneva, Switzerland (2000)
5. ISO 12572. Hygrothermal Performance of Building Materials and Products—Determination of water vapour transmission properties, Geneva, Switzerland (2001)
6. ISO 12571. Hygrothermal performance of building materials and products—Determination of hygroscopic sorption properties. Geneva. Switzerland: International Organization for Standardization (2000)
7. EN 12859. Gypsum blocks. Definitions, requirements and test methods, Geneva, Switzerland (2011)
8. ISO 15148. Hygrothermal Performance of Building Materials and Products. Determination of Water Absorption Coefficient by Partial Immersion. Geneva. Switzerland: International Organization for Standardization (2002)
9. CEN/TC 89. Thermal Performance of Buildings and Building Components—Determination of Water Absorption Coefficient, Geneva, Switzerland (1994)
10. M. Bomberg, *Moisture Flow Through Porous Building Materials* (Division of Building Technology, Lund Institut of Technology, Lund, Sweden, Report n° 52, 1974)
11. H. Janssen, H. Derluyn, J. Carmeliet, Moisture transfer through mortar joints: a sharp-front analysis. Cement Concrete Res. **42**, 1105–1112 (2012)
12. P. Mukhopadhyaya, P. Goudreau, K. Kumaran, N. Normandin, Effect of surface temperature on water absorption coefficient of building materials. J. Therm. Envelope Build. Sci. **26**, 179–195 (2002)
13. S. Roels et al., Interlaboratory comparison of hygric properties of porous building materials. J. Therm. Environ. Build. **27**(4), 307–325 (2004)
14. N.A. García et al., Propiedades físicas y mecánicas de ladrillos macizos cerámicos para mampostería. Ciencia y Ingeniería Neogranadina **22**(1), 43–58 (2012)
15. J. Azevedo, Absorção por capilaridade de soluções Salinas em materiais porosos, M.Sc. thesis, Faculdade de Engenharia da Universidade do Porto, Portugal (2013)
16. A.S. Guimarães, J.M.P.Q. Delgado, T. Rego, V.P. De Freitas, The effect of salt solutions in the capillarity absorption coefficient of red brick samples. Defect Diffusion Forum **369**, 168–172 (2016)
17. J. Crank, *The Mathematics of Diffusion* (Oxford Science, New York, USA, 1975)
18. V.P. Freitas, Moisture transfer in building walls – Interface phenomenon analyse, Ph.D. Thesis, Faculty of Engineering, University of Porto, Porto, Portugal (1992)
19. P. Crausse, Fundamental study of heat and moisture transfer in unsaturated porous medium, Ph.D. thesis, ENSEEIHT, Toulouse, France (1983)
20. B. Perrin, Study of coupled heat and mass transfer in consolidated, unsaturated porous materials used in civil engineering, Ph.D. Thesis, University Paul Sabatier (1985)
21. V. Koci et al., Heat and moisture transport and storage parameters of bricks affected by the environment. Int. J. Thermophys. **39**, 63 (2018)
22. ISO 8301. Thermal insulation: Determination of steady-state thermal resistance and related properties—Heat flow meter apparatus. Geneva, Switzerland: International Organization for Standardization (1991)
23. D. Taoukil, A. El Bouardi, F. Sick, A. Mimet, H. Ezbakhe, T. Ajzoul, Moisture content influence on the thermal conductivity and diffusivity of wood–concrete composite. Constr. Build. Mater. **48**, 104–115 (2013)

24. NBR 7200. Execution of Wall and Ceiling Coatings of Inorganic Mortars—Procedure, Rio de Janeiro, Brazil (1982)
25. ASTM C 270. Standard Specification for Mortar for Unit Masonry, Revision 14A (2014)
26. NBR NM 248. Aggregates—Sieve analysis of fine and coarse aggregates, Rio de Janeiro, Brazil (2009)
27. NBR NM 52. Fine aggregate—Determination of the bulk specific gravity and apparent specific gravity, Rio de Janeiro, Brazil (2009)
28. NBR NM 46. Aggregates—Determination of material finer than 75 μm sieve by washing, Rio de Janeiro, Brazil (2003)
29. NBR NM 45. Aggregates—Determination of the unit weight and air-void contents, Rio de Janeiro, Brazil (2006)
30. NBR NM 30. Fine aggregate—Test method for water absorption, Rio de Janeiro, Brazil (2001)
31. NBR 7211. Aggregates for concrete—Specification, Rio de Janeiro, Brazil (2009)
32. NBR 11579. Portland cement—Determination of fineness index by means of the 75 μm sieve (n° 200), Rio de Janeiro, Brazil (2012)
33. NBR 16372. Portland cement and other powdered materials—Determination of fineness by the air permeability method (Blaine method), Rio de Janeiro, Brazil (2015)
34. NBR NM 65. Portland cement—Determination of setting times, Rio de Janeiro, Brazil (2003)
35. NBR 11582. Portland cement—Determination of soundness by Le Chatelier test, Rio de Janeiro, Brazil (2016)
36. NBR 7215. Portland cement—Determination of compressive strength, Rio de Janeiro, Brazil (1996)
37. C. Nunes, Z. Slížková, Hydrophobic lime based mortars with linseed oil: characterization and durability assessment. Cem. Concr. Res. **61**, 28–39 (2014)
38. NBR 13276. Mortars applied on walls and ceilings—Preparation of mortar for unit masonry and rendering with standard consistence index, Rio de Janeiro, Brazil (2005)
39. NBR 13278. Mortar—Determination of the specific gravity and the air entrained content in the fresh stage—Method of test, Rio de Janeiro, Brazil (2005)
40. NBR 13279. Mortars applied on wall and ceilings—Determination of the flexural and the compressive strength in the hardened stage, Rio de Janeiro, Brazil (2005)
41. J. Fort, M. Čáchová, E. Vejmelková, R. Černy, Mechanical and hygric properties of lime plasters modified by biomass fly ash, in *IOP Conference Series Materials Science and Engineering* (2018), p. 365
42. E. Vejmelková, M. Keppert, M Jerman, R. Černy, Basic physical, mechanical and hygric properties of renders suitable for historical buildings, in *IOP Conference Series Materials Science and Engineering* (2018), p. 364
43. S.K. Ghosh, M. Melander, *Air content of mortar and water penetration of masonry walls. Masonry information* (Portland Cement Association, Skokie, 1991)
44. N.L. Mustelier, Estimativa do comportamento de paredes no ensaio de penetração de água de chuva através das propriedades de transferência de humidade dos materiais constituintes, Ph.D. thesis, Centro Tecnológico da Universidade Federal de Santa Catarina, Brazil (2008)
45. J.F. Daians, Processus de condensation et de transfert d'eau dans un matériau meso et macroporeux: étude expérimentale du mortier de ciment, Ph.D. thesis, Université Scientifique et Médicale de Grenoble (1986)
46. C.P. Fernandes, et al. Contribuição ao Estudo da Migração da água em Materiais Porosos Consolidados: Análise de Uma Argamassa de Cal e Cimento, in *Proceedings of the 10th Brazilian Congress of Mechanical Engineering* (ABCM, Rio de Janeiro, Brazil, 1989), pp. 557–560
47. F.M. Fernandes, P.B. Lourenço. Estado da arte sobre tijolos antigos. In Construção 2007, pp. 27–33, Coimbra, Portugal, 17–19 December (2007)
48. C. Nunes, L. Pel, J. Kunecký, Z. Slížková, The influence of the pore structure on the moisture transport in lime plaster-brick systems as studied by NMR. Constr. Build. Mater. **142**, 395–409 (2017)
49. H. Künzel, *Simultaneous Heat and Moisture Transport in Building Components—One and two-dimensional calculation using simple parameters* (IBP Verlag. Stuttgart, Germany, 1995)

50. H. Derluyn, H. Janssen, J. Carmeliet, Influence of the nature of interfaces on the capillary transport in layered materials. Constr. Build. Mater. **25**, 3685–3693 (2011)
51. N. Laaroussi et al., Measurement of thermal properties of brick materials based on clay mixtures. Constr. Build. Mater. **70**, 351–361 (2014)
52. Z. Suchorab, D. Barnat-Hunek, H. Sobczuk, Influence of moisture on heat conductivity coefficient of aerated concrete. Ecol. Chem. Eng. **18**, 111–120 (2011)
53. T.S.M.R. Rêgo, Efeito de Soluções Aquosas Salinas nos Processos de Embebição de Paredes com Múltiplas Camadas: Faculdade de Engenharia da Universidade do. M.Sc., thesis, Porto (2014)
54. T.D. Gonçalves, V. Brito, L. Pel, Water vapor emission from rigid mesoporous materials during the constant drying rate period. Drying Technol. **30**(5), 462–474 (2012)
55. W. Depraetere, J. Carmeliet, H. Hens, Moisture transfer at interfaces of porous materials: measurements and simulations. PRO **12**, 249–259 (2000)
56. V. Corinaldesi et al., Mechanical behaviour and thermal conductivity of mortars containing waste rubber particles. Mater. Des. **32**, 1646–1650 (2011)
57. H. Carasek, O. Cascudo, L.M. Scartezini, Importância dos materiais na aderência dos revestimentos de argamassa. Simpósio Brasileiro de Tecnologia das Argamassas, p. 43–67 (2001)
58. R. Demirboğa, R. Gül, The effects of expanded perlite aggregate, silica fume and fly ash on the thermal conductivity of lightweight concrete. Cem. Concr. Res. **33**(5), 723–727 (2003)
59. Y. Xu, D.D.L. Chung, Effect of sand addition on the specific heat and thermal conductivity of cement. Cem. Concr. Res. **30**, 59–61 (2000)
60. M.A. Wilson, W.D. Hoff, C. Hall, Water movement in porous building materials—XIII absorption in two—layer composite. Build. Environ. **30**, 209–219 (1995)

Chapter 3
Interface Influence During the Drying Process

3.1 Introduction

The study of drying kinetics of building materials is very relevant to avoid damage in the construction industry, taking into account that the drying process plays an important role in the available humidity, inside the material and on its surface. Drying can be defined as the process by which water leaves a porous building material, knowledge and understanding of which is necessary to predict the performance of those materials in service.

In this experimental study, the interface influence on the drying process of multilayer ceramic blocks with perfect contact, air space and hydraulic contact (cement and lime mortar) at different heights was analysed. The results showed an increase in the drying time constant for the materials with interface compared to monolithic materials. In addition, the farther away the interface is located from the base, the longer the drying time constant. The interface significantly retarded the flow of moisture transport. i.e., the discontinuity of moisture content across the interface, which indicated that capillary transport, deviates from the classical capillary transport in a homogeneous material.

Several empirical and semi-empirical models are presented, in literature, to describe the drying process of porous materials, especially for food and chemical industries [1–3], where this phenomenon is fundamental.

Related to the field of civil engineering, several authors have studied multi-layered walls mainly with perfect contact interfaces [4–6]. The hypothesis of perfect contact implies, in most cases, that the interface will be considered to have no effect on moisture transport [7]. However, related to models with imperfect contact, literature is less abundant. Freitas et al. [8] studied the interface influence on the drying kinetics of samples of cellular concrete and red brick. Derdour et al. [9] analysed the effect of thickness, porosity and the drying conditions of several building materials on the drying time constant. The authors concluded that the relationships corresponding to

J. M. P. Q. Delgado et al., *Drying Kinetics in Building Materials and Components*, SpringerBriefs in Applied Sciences and Technology, https://doi.org/10.1007/978-3-030-31860-4_3

the characteristic drying curve obtained for thick components depend on the initial conditions. On the other hand, they observed a strongly dependence of the thicknesses and various plaster textures on the characteristic drying curve obtained. Bednar [10] studied the influence of the following properties: material size and insulation, and climate conditions on the liquid moisture transport coefficient during drying experiments. Karoglou et al. [11] evaluated the effect of environmental conditions, such as air temperature, air relative humidity and air velocity, on the drying performance of 4 stone materials, 2 bricks and 7 plasters. The authors showed that the parameters of the model proposed were affected strongly by the material and the drying air conditions.

The transfer of moisture in buildings is usually modelled using a "diffusion" approach, because the different transport mechanisms are combined into a single moisture diffusivity that depends on the moisture content. In recent years several methods have been presented in the literature [12] to simplify the determination of moisture diffusivity from measurements of the water profile content (using the gamma or nuclear magnetic resonance (NMR) method), such as the Boltzmann transform or the flow gradient method.

In this chapter, the influence on the drying process of ceramic blocks, with different interfaces types at different heights, is analysed. The variation observed in the drying time constants for the materials with interface (at different interface heights), comparatively to the monolithic materials, is presented, as the influence of each interface type.

3.2   Theory: Drying Kinetics and Modelling

Thin-layer equations can be classified into empirical, semi-empirical and theoretical equations [13]. These are generally linear, power, exponential, Arrhenius or Logarithmic types. A common semi-empirical equation that describes the drying process is:

$$MR = \frac{w - w_\infty}{w_0 - w_\infty} = \exp^{-t^n/t_c} \qquad (3.1)$$

where MR is the moisture ratio (dimensional), w is the total moisture content (in kg/m^3) is the moisture content, w_∞ is the equilibrium moisture content (in kg/m^3) after the drying process, w_0 (kg/kg) is the initial moisture content in the beginning of the drying process (in kg/m^3), t is the time (h), n is the empirical coefficient of model (dimensional) and t_c is the drying time constant (h) ($K = 1/t_c$ where K (h^{-1}) is the drying constant).

Over the last years, it has been possible to find in literature several studies carried out related to drying kinetics [14–19]. The first kinetic model developed was the Newton model. Equation (3.1) with $n = 1$, obtained by simplifying Fick's second law. Through the Newton model, studies have emerged where new models have been

Table 3.1 Thin layer drying models tested for modelling

Model name	Model equation
Newton [14, 15]	$MR = e^{-t/t_c}$
Page [14, 15]	$MR = e^{-t^n/t_c}$
Modified Page [14]	$MR = e^{-(t/t_c)^n}$
Henderson and Pabis [14, 15]	$MR = a \cdot e^{-t/t_c}$
Two-term model [14, 15]	$MR = a \cdot e^{-t/t_{c1}} + b \cdot e^{-t/t_{c2}}$
Diffusion approach or Two-term exponential [14, 15]	$MR = a \cdot e^{-t/t_{c1}} + (1-a) \cdot e^{-t/t_{c2}}$
Henderson and Pabis modified [18]	$MR = a \cdot e^{-t/t_{c1}} + b \cdot e^{-t/t_{c2}} + c \cdot e^{-t/t_{c3}}$
Logarithmic [14, 15]	$MR = a \cdot e^{-t/t_c} + c$
Midilli-Kucuk [15]	$MR = a \cdot e^{-t^n/t_c} + b \cdot t$
Thompson [17]	$t = a \ln(MR) + b[\ln(MR)]^2$
Wang and Singh [14]	$MR = 1 + a \cdot t + b \cdot t^2$
Weibull distribution [18]	$MR = e^{-(t,t_c)^n}$
Aghbashlo et al. [16]	$MR = e^{(t/t_{c1}/(1+t/t_{c2}))}$
Verma et al. [14, 15]	$MR = a \cdot e^{-t/t_{c1}} + (1-a)e^{-t/t_{c2}}$
Vega and Lemus [19]	$MR = (a + t/t_c)^2$
Hii et al. [14, 15]	$MR = a \cdot e^{-(t^n/t_{c1})} + b \cdot e^{-(t^n/t_{c2})}$

designed for different tested materials, as described in Table 3.1 which presents an exhaustive list of empirical and semi-empirical models found in the literature.

The experimental drying curves presented in this study were fitted with the models described in Table 3.1 and regression analyses of these equations were done by using STATISTICA routine (see Tables 3.2, 3.3 and 3.4 the most suitable approximation). The regression coefficient (R^2) was the primary criterion for selecting the best equation to describe the drying curves of tested materials. The performance of derived new models was evaluated using different statistical parameters such as the mean bias error (*MBE*), the root mean square error (*RMSE*) and chi-square (χ^2), in addition to the regression coefficient (R^2). These parameters can be calculated as following:

$$MBE = \frac{1}{N} \sum_{i=1}^{N} \left(MR_{pre,i} - MR_{\exp,i} \right) \tag{3.2a}$$

$$RMSE = \sqrt{\frac{1}{N} \sum_{i=1}^{N} \left(MR_{pre,i} - MR_{\exp,i} \right)} \tag{3.2b}$$

$$\chi^2 = \frac{\sum_{i=1}^{N} \left(MR_{\exp,i} - MR_{pre,i} \right)^2}{N - n} \tag{3.2c}$$

where *MR* is the moisture ratio.

Table 3.2 Results of the statistical analyses obtained with several tested drying models

Sample		T (°C)	RH (%)	Error (%)					
				Newton	Page	Page Mod	Logarit.	Midilli et al.	Hend.-Pabis
A	Monolithic	20	50	7.8	4.5	7.8	0.3	1.0	7.4
A	PC at 2 cm	20	50	3.8	3.7	3.8	0.4	0.7	3.5
A	PC at 5 cm	20	50	7.2	2.7	7.1	0.2	0.5	6.1
A	PC at 7 cm	20	50	10.5	2.7	10.4	0.3	0.5	8.1
B	Monolithic	20	50	1.8	3.9	2.6	0.6	2.8	2.2
B	PC at 2 cm	20	50	3.9	7.5	4.6	1.2	4.6	4.8
B	PC at 5 cm	20	50	23.8	16.4	31.9	12.7	12.7	28.8
B	PC at 7 cm	20	50	81.1	70.4	102.7	0.4	66.4	93.8
Sum				141.9	111.8	172.7	16.3	89.5	156.9

Table 3.3 Values of drying constant obtained for the best fitted models

Sample		T (°C)	RH (%)	K (h^{-1})					
				Newton	Page	Page Mod	Logarit.	Midilli et al.	Hend.-Pabis
A	Monolithic	20	50	0.004	0.012	0.004	**0.006**	0.009	0.004
A	PC at 2 cm	20	50	0.004	0.009	0.004	**0.005**	0.005	0.004
A	PC at 5 cm	20	50	0.004	0.012	0.004	**0.005**	0.008	0.003
A	PC at 7 cm	20	50	0.003	0.013	0.003	**0.005**	0.009	0.003
B	Monolithic	20	50	0.003	0.002	0.003	*0.004*	0.001	0.003
B	PC at 2 cm	20	50	0.003	0.002	0.003	*0.004*	0.001	0.003
B	PC at 5 cm	20	50	0.002	0.007	0.003	*0.003*	0.003	0.003
B	PC at 7 cm	20	50	0.002	0.009	0.003	*0.002*	0.009	0.002

In bold and italics the best results obtained

Table 3.4 Values of drying time constant obtained for the best fitted models

Sample		T (°C)	RH (%)	t_c (h)					
				Newton	Page	Page Mod	Logarit.	Midilli et al.	Hend.-Pabis
A	Monolithic	20	50	224.9	84.1	224.9	**161.0**	116.4	236.3
A	PC at 2 cm	20	50	247.2	110.0	247.2	**183.6**	184.0	258.6
A	PC at 5 cm	20	50	264.6	86.9	264.6	**191.9**	124.2	287.5
A	PC at 7 cm	20	50	303.5	74.1	303.5	**202.6**	114.8	349.0
B	Monolithic	20	50	331.3	556.1	323.8	267.2	958.0	312.8
B	PC at 2 cm	20	50	353.6	547.3	340.5	271.6	1231.9	336.0
B	PC at 5 cm	20	50	408.2	138.3	347.5	*312.5*	329.4	363.9
B	PC at 7 cm	20	50	418.9	110.0	383.8	*416.0*	107.4	421.1

In bold and italics the best results obtained

3.3 Results and Discussion

3.3.1 Drying Experiments/Drying Kinetics

The experiments to determine the interface influence in ceramic blocks on the drying kinetics were performed at constant conditions of relative humidity and temperature. For the drying test, the specimens were immersed in a recipient containing water where they stayed until reaching the saturation point. To measure the drying time of the specimens it was carried out two different environments for the specimens containing mortar interface. Only one type of interface was chosen, with the goal of to verify that when the drying speed increases there is an increase in the resistance of the interface. First, these specimens were exposed to environments under temperatures of 20 °C and 50% RH and later were placed inside a hermetic box with a temperature of 70 °C and 3% RH. The specimens with air space interface and perfect contact were subjected only to the first environmental conditions. It is important to highlight that between the first and the second environmental conditions, the specimens with hydraulic contact interface were immersed inside a recipient filled with water up to the saturation point again.

The tested samples had an approximate size of $4.0 \times 4.0 \times 10 \, cm^3$ (ceramic block A) and $3.0 \times 3.0 \times 10 \, cm^3$ (ceramic block B) as it was made in Chap. 4 for the absorption tests. The samples' faces were completely sealed, with an epoxy material, except the upper face (see Fig. 3.1).

The samples used in the drying process were the same ones used in the absorption phase, which only needed the waterproofing on the underside (Fig. 3.1). In this study (drying kinetics analysis), three different interface heights were analysed (at 2 cm from the base, at 5 cm, and at 7 cm).

To perform these experiments, a climatic chamber was used to guarantee the adequate climatic conditions over extended periods of time, namely to control the temperature and relative humidity. Both parameters could be independently set to constant values. A precision balance was located inside the climatic chamber and the mass change registered continuously by a computer.

The drying semi-empirical models presented in Table 3.1 were used to describe drying curves and the experimental results obtained for the ceramic block tested are presented in Table 3.2. The semi-empirical models selected to drying analysis were the ones who gathered better results. The criterion used for selecting the models that better describe the drying process was the magnitude of the relative error for each one (see Eq. 3.2).

According to the results of Table 3.2, the logarithmic model presented the best results (see the errors sum). Tables 3.3 and 3.4 show the drying time constant and the drying constant obtained.

Figure 3.2 (for ceramic block A) and Fig. 3.3 (for ceramic block B) show the changes in the moisture ratio (*MR*) with time during the drying process, at 20 °C and $RH = 50\%$, for monolithic and perfect contact. It is also possible to see the flux along the time for the same cases.

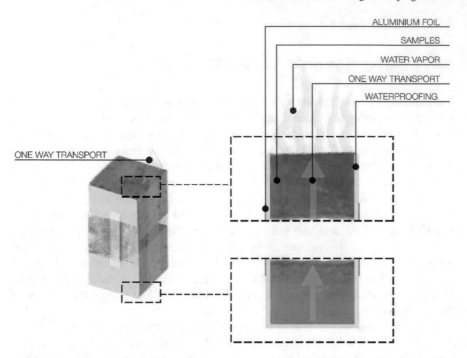

Fig. 3.1 Drying models tested

As expected, the results show an increase in drying times for materials with perfect contact interface, compared to the monolithic materials; and show that the further away from the base the interface is located, the greater is the drying time constant.

To more fully describe the normalization adopted for the following figures, it is presented, in Tables 3.5 and 3.6, for the weighing difference and HUMIGAMA tests, the procedure adopted. Normalizing the graphs presented by Figs. 3.4, 3.5, 3.6 and 3.7, it was possible to compare the two methods studied (HUMIGAMA and weighing difference).

From Figs. 3.4, 3.5, 3.6 and 3.7 show similarity in the results obtained, validating the two methods and ensuring that there is an increase in the drying time constant for interfaced materials compared to monolithic materials, and the further away from the base the interface is located, the longer the time of constant drying process. In resume, the interface could significantly retard the moisture transport, i.e., the discontinuity of moisture content across the interface indicated that there was a difference in capillary pressure across the interface.

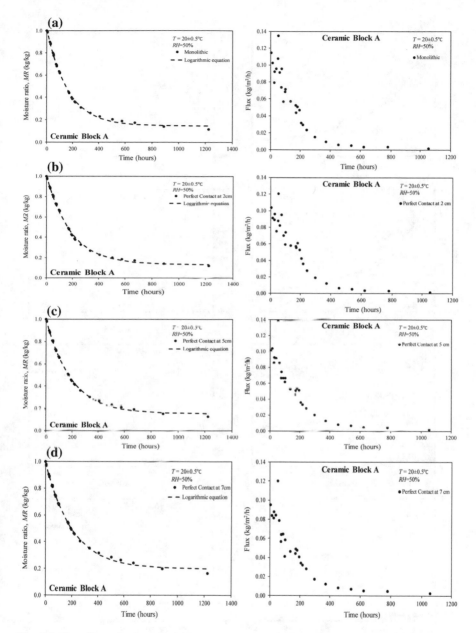

Fig. 3.2 Experimental values of the moisture content and flux mass, of ceramic block A, versus drying time, for: **a** monolithic sample; **b** sample with perfect contact at 2 cm; **c** sample with perfect contact at 5 cm and **d** sample with perfect contact at 7 cm

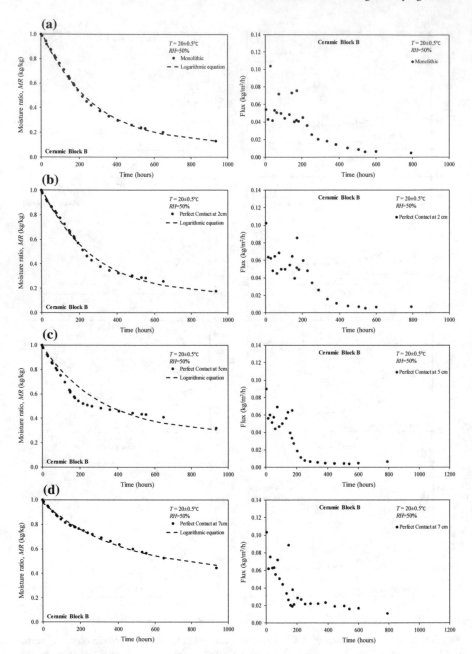

Fig. 3.3 Experimental values of the moisture content and flux mass, of ceramic block B, versus drying time, for: **a** monolithic sample; **b** sample with perfect contact at 2 cm; **c** sample with perfect contact at 5 cm and **d** sample with perfect contact at 7 cm

Table 3.5 Example of normalization for the calculation of *MR* by weighing difference

W (the total moisture content) (kg)	Time (hours)	Moisture ratio ((kg/kg), $MR = (w - w_\infty)/(w_0 - w_\infty)$
0.356235	0.0	1.00
0.355947	3.0	0.99
0.355544	3.0	0.98
…	…	…
0.323079	672.0	0.17
0.321753	888.0	0.14
0.320784	1224.0	0.12
The initial moisture content in the beginning of the drying process (w_0) (kg)		0.35624
The equilibrium moisture content after drying process (w_∞) (kg)		0.31616

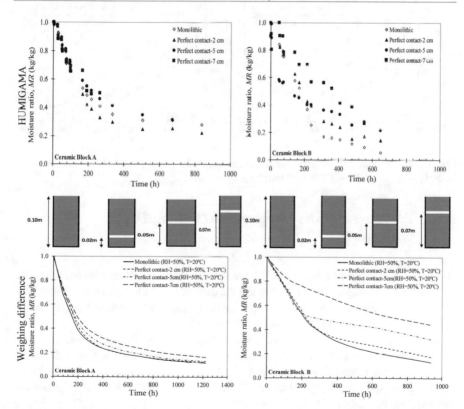

Fig. 3.4 Effect of perfect contact interface on moisture content for ceramic block A and ceramic block B, comparing of the HUMIGAMA versus weighing test

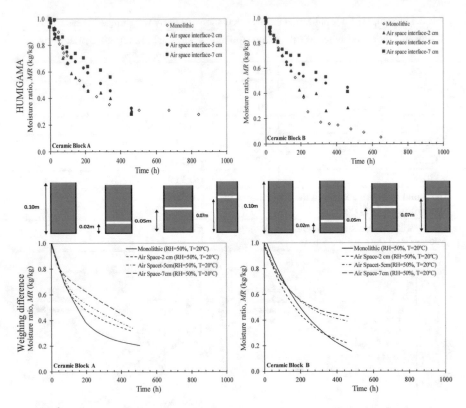

Fig. 3.5 Effect of air space interface on moisture content for ceramic block A and ceramic block B, comparing of the HUMIGAMA vs weighing test

3.3.2 Moisture Profiles Comparison (Wetting/Drying Processes)

HUMIGAMA.VF has as one of the advantages to obtain moisture profiles along the thickness of the sample allowing knowing the distribution of the moisture content, for this purpose the monolithic samples, comparing these with samples having perfect contact, air space and hydraulic contact interface.

Figure 3.8 presents the moisture content profiles along the specimen thickness for the monolithic samples tested with gamma ray attenuation in drying and wetting processes. This figure shows that in a single piece of ceramic block the wetting process (used only for comparison, see Guimarães et al. [20]) and the drying process present different behaviors, that is, during the drying process it is possible to observe a homogeneous drying. More precisely, the moisture profiles become progressively flatter as the drying proceeds, i.e., the moisture content is uniformly distributed over the specimen thickness and is homogeneously decreased. In addition, since no evaporation front is present, it can be assumed that this experiment corresponds to the

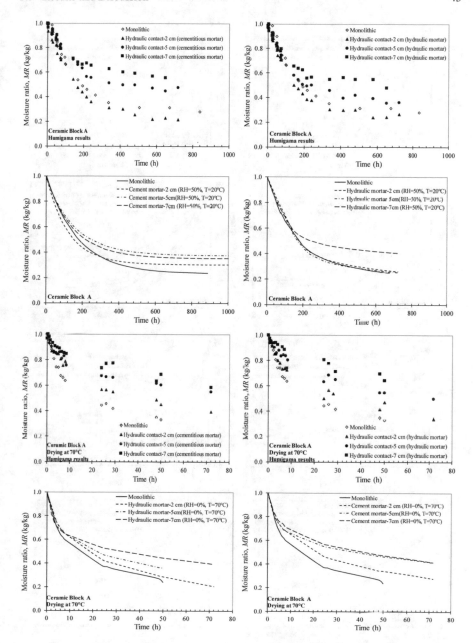

Fig. 3.6 Effect of hydraulic lime mortar and cement mortar interface on moisture in ceramic block A, comparing of the HUMIGAMA vs weighing test

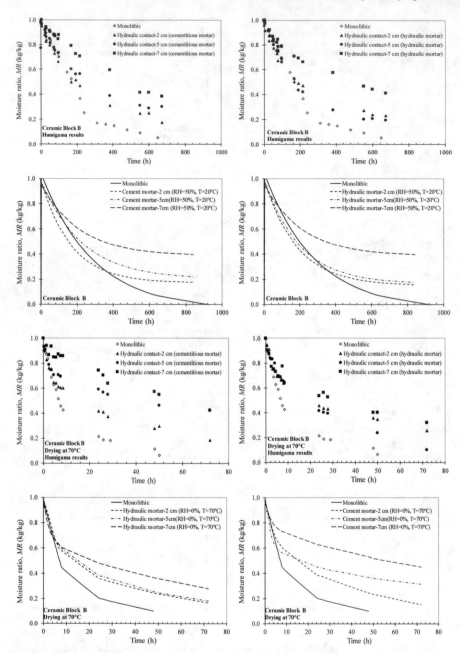

Fig. 3.7 Effect of hydraulic lime mortar and cement mortar interface on moisture in ceramic block B, comparing of the HUMIGAMA vs weighing test

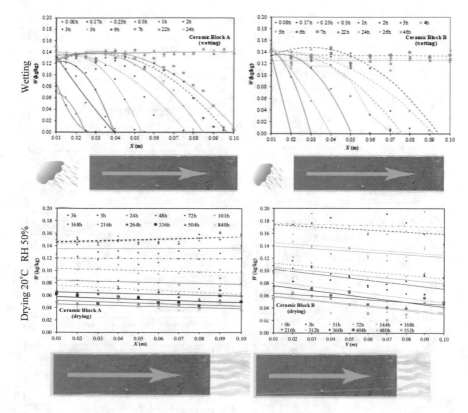

Fig. 3.8 Moisture content of monolithic samples (ceramic block A and B), during the wetting and drying processes

1st drying phase and that moisture is transferred by capillary forces [21]. Evaporation occurs first in the larger pores and then in the smaller ones.

3.3.3 Interface Influence

3.3.3.1 Perfect Contact

Figures 3.9 and 3.10 show some examples of the perfect contact interface influence on the wetting and drying processes. As for monolithic samples, the moisture content is evenly distributed over the specimen thickness and decreases homogeneously. As observed in Sect. 3.3.2. the gamma ray results showed an increase in the drying time constant for the materials with perfect contact interface comparatively to the monolithic materials, and that the further away from the base the interface is located, the longer the drying time constant [20].

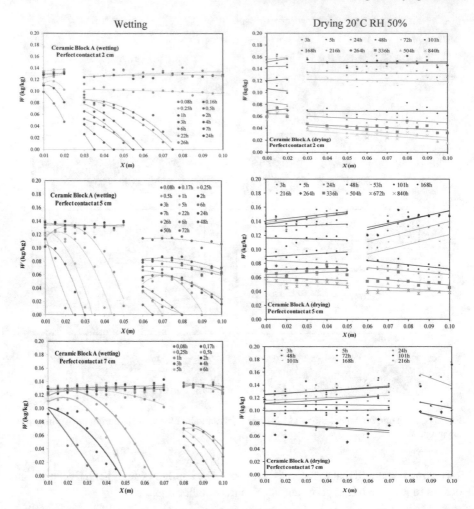

Fig. 3.9 Moisture content along the thickness, for samples of ceramic block A, with perfect contact, during the wetting and drying processes, at 2, 5 and 7 cm

3.3.3.2 Air Space Interface

In air space, its interface influence on wetting and drying processes will be illustrated from Figs. 3.11 and 3.12. The moisture content is evenly distributed over the specimen thickness and decreases homogeneously, like in monolithic samples. Compared to this kind of sample (monolithic), the results obtained in Sect. 3.3.2. (gamma ray) presented an increase of the drying time constant for the materials with air space, which means that the drying process is more slowly with air space interface. Moreover, the further the interface is from the base, the more the constant drying time increases [20].

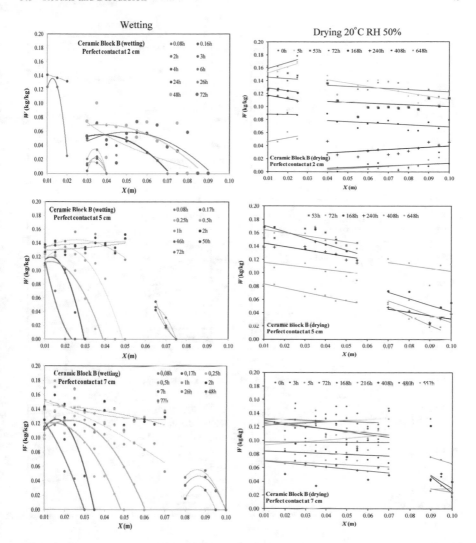

Fig. 3.10 Moisture content along the thickness, for samples of ceramic block B, with perfect contact, during the wetting and drying processes, at 2, 5 and 7 cm

3.3.3.3 Hydraulic Contact—Cementitious Mortar

As shown by Nunes et al. [22, 23], the changing of porosity in mortar near the brick-mortar transition zone can be attributed to the flow of water from fresh brick during the bonding process. In the beginning, small binder particles can be transported to the mortar-brick interface and the plaster becomes more compact (enriched in the binder agent) in the interface [23–25] studied the phenomenon of water from a cement-to-brick grout and water retention of the plaster, i.e. the amount of water retained by

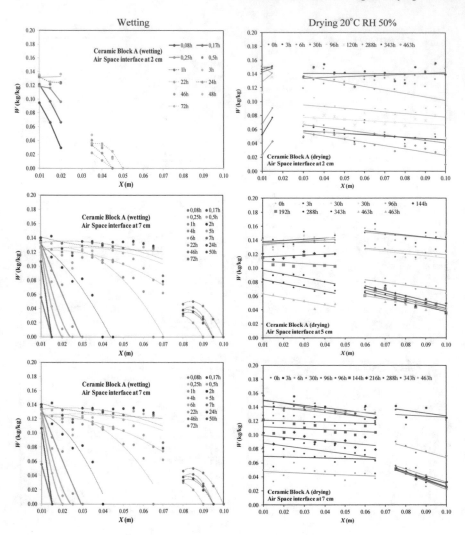

Fig. 3.11 Ceramic block A, with air space interface, at 2, 5 and 7 cm

fresh plaster. Figures 3.13 and 3.14 present the hydraulic contact interface influence
in regard to wetting and drying processes. Such as previously mentioned monolithic
samples, the moisture content is equally distributed over the specimen thickness
and decreases homogeneously. As mentioned before, the gamma ray results of the
materials with hydraulic contact interface showed an increase in the drying time
constant, when compared to the monolithic materials, and that the further away from
the base the interface is located, the greater the constant drying time [20].

A detailed analysis was done and it is based on the comparison of the average
moisture content over the sample thickness before (bottom) and after (upper) the

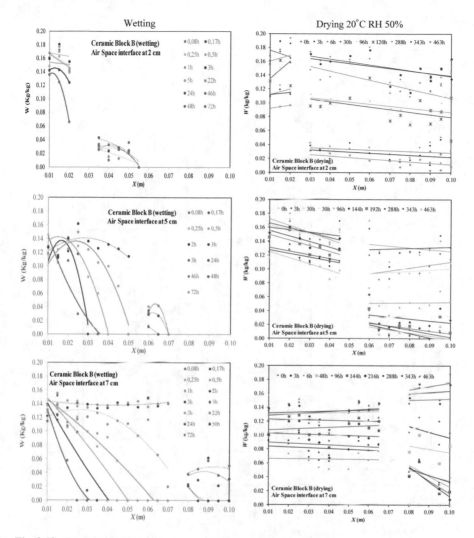

Fig. 3.12 Ceramic block B, with air space interface, at 2, 5 and 7 cm

interface, at the beginning and at the end of the experimental test. The results obtained are presented in Tables 3.5 and 3.6.

When comparing the profiles submitted to drying in an environment with 20 °C and 50% RH, with cement mortar hydraulic contact interface can be observed:

- **Ceramics "A":**

At the outset of specimen testing, both sides of the test specimen (upper and lower) had 0.14 kg/kg of water. For specimens with an interface at 2 cm from the base with sides both dried simultaneously over 720 h, reached 0.06 kg/kg at the bottom and 0.02 kg/kg at the top, different from what happened with the interface sample at

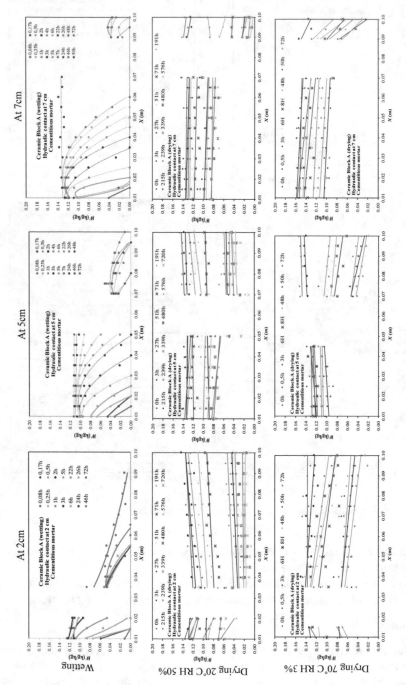

Fig. 3.13 Ceramic block A, with hydraulic contact of cementitious mortar, at 2, 5 and 7 cm

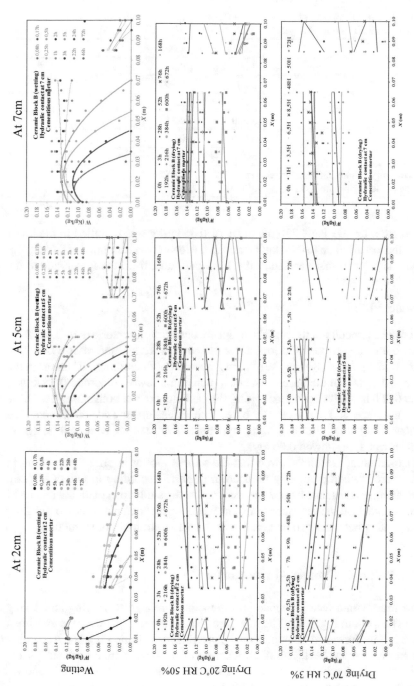

Fig. 3.14 Ceramic block B, with hydraulic contact of cementitious mortar, at 2, 5 and 7 cm

Table 3.6 Example of normalization for the calculation of MR by HUMIGAMA

W_{med} (average of the amount of moisture in the entire sample) (kg/kg)	Time (hours)	Moisture ratio ((kg/kg), $MR_{med} = W_{med}/W_{max}$
0.151	3.0	1.00
0.150	3.0	0.99
0.149	24.0	0.99
...	...	...
0.047	504.0	0.31
0.047	672.0	0.31
0.042	840.0	0.28
The first average amount of moisture content in the beginning of the drying process (w_{max}) (kg/kg)		0.151

5 cm. Where after the same 720 h a delay of 0.06 kg/kg was recorded in the lower part, i.e., the top dried up to 0.04 kg/kg, and in the lower part it stopped at 0.08 kg/kg. Behaviour similar to that verified at the 7 cm interface, from the lower side to the upper side, over 648 h.

- **Ceramics "B":**

At the outset of specimen testing, those specimens with the interface at 2 cm from the base, both sides (upper and lower) registered 0.14 kg/kg, and after 672 h both sides had 0.02 kg/kg, presenting similar behaviour at both extremes. Those with the interface at 5 cm from the base, both sides (upper and lower) recorded 0.15 kg/kg, and after 672 h had 0.04 kg/kg, exhibiting similar behaviour at both extremes. Those with the interface at 7 cm from the base, both sides (upper and lower) recorded 0.14 kg/kg which, and after 672 h, the upper side recorded an average 0.02 kg/kg, while the lower side showed 0.06 kg/kg.

The experimental results showed that the cement mortar interface may result in significant error when predicting moisture transport. According to Qiu et al. [26], the mismatching resistance hygric was assumed in the study to explain the impact of hydraulic contact on the moisture transport while perfect contact refers to the good physical contact between two materials without penetration of pore structure.

3.3.3.4   Hydraulic Contact—Lime Mortar

The lime mortar interface influence on the methods related to wetting and drying phase can be seen in Fig. 3.15. The moisture content distribution over the specimen thickness is a uniform and homogeneously, just like monolithic samples. According to gamma ray results (Sect. 2.4.2), an increase of the drying time constant for the materials with hydraulic lime mortar, parallel to monolithic materials, is shown. Besides that, the further from the base the interface is located, more constant drying time increases (Fig. 3.16) [20].

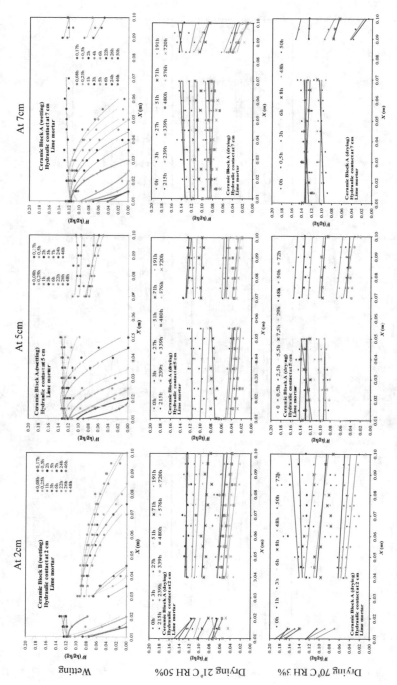

Fig. 3.15 Ceramic block A, with hydraulic contact of hydraulic mortar, at 2, 5 and 7 cm

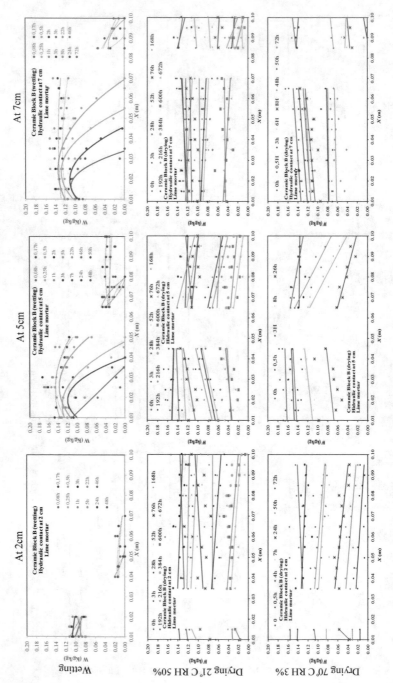

Fig. 3.16 Ceramic block B, with hydraulic contact of hydraulic mortar, at 2 cm, 5 cm and 7 cm

When comparing the drying profiles submitted in an environment of 20 °C and 50% RH with lime mortar hydraulic contact interface, can be observes:

- **Ceramics "A":**

At the outset of specimen testing, those specimens having the interface at 2 cm from the base, both sides (upper and lower) registered 0.13 kg/kg and after 720 h the lower and upper side registered 0.05 kg/kg. Those with the interface at 5 cm from the base, both sides (upper and lower) registered 0.12 kg/kg, and after 720 h both sides had on average 0.03 kg/kg, presenting similar behavior at both extremes. Those with the interface at 7 cm from the base, both sides (upper and lower) registered 0.14 kg/kg and after 720 h showed that on the lower side there was an average of 0.08 kg/kg, while the upper one showed 0.02 kg/kg. This indicates that the 7 cm interface created greater water resistance in the transport of moisture from the lower side to the upper side.

- **For ceramics "B":**

At the outset of specimen testing, those with the interface at 2 cm from the base, both sides (upper and lower) recorded 0.14 kg/kg and after 672 h both sides had 0.2 kg/kg. Those having the interface at 5 cm from the base, the lower side registered 0.15 kg/kg and the upper side 0.14 kg/kg, and after 672 h registered, the lower side 0.03 kg/kg, and the upper 0.02 kg/kg. Those with the interface at 7 cm from the base, both sides (upper and lower) registered an average of 0.13 kg/kg, and after 672 h showed that on the lower side there was an average of 0.04 kg/kg, while the upper one showed 0.01 kg/kg. As it was said the 0.7 cm interface created greater water resistance in the transport of moisture in relation to the interfaces at 5 and 2 cm.

It has been found that, in the drying process, when the flow reached the interface is greater than the maximum flow transmitted, there is a water resistance and consequently a faster drying of the outer layer due to lack of water supply and a delay in the drying of the inner layer. These results were observed in the environment of 20 °C and 50% RH only in the specimens with interface at 7 cm, as well as in 10 of the 12 cases in an environment of 70 °C and 3% RH. This corroborates the prediction made by Freitas [8], where if the drying takes place naturally, i.e. in the hygrothermal conditions existing in the laboratory itself, the flows involved are reduced and, therefore, the potential flows that seek to cross the interface are always inferior to the maximum flow transmitted through the interface, with no alteration in the development of the water profiles.

The gamma ray technique allows the analysis of the sample in an accurate position, which is extremely advantageous when studying layered materials. The moisture profiles showed that the lime mortar inhibited the brick drying slightly, unlike the wetting process where the mortar has a significant moisture resistance. Since the monolithic brick samples showed similar drying kinetics between the samples with cement mortar interface and lime mortar, the main reason for the remarkable reduction in the drying rate of the brick system seems to be related mainly to a hydraulic resistivity created at the interface between the plaster and the brick. Thus, the combination of mortars with substrates is of great importance in the study of drying kinetics of a layered system. When plaster is applied and cured on a porous substrate, physical contact between two materials with penetration of pore structure, the porous network changes along the specimen thickness and, in particular, at the interface, which influences the drying behaviour of the entire system.

3.4 Synthesis

This section presents an experimental study to analyse the interface influence on the drying process of ceramic blocks with different interfaces (perfect contact, hydraulic contact and air space), at different interface highs (2, 5 and 7 cm). The main achievements and conclusions drawn from this study are as follows:

- The study of drying kinetics in building materials is very relevant to the construction industry in order to avoid building damage. Moisture damage is one of the most important factors limiting building performance, especially for its durability, waterproofing, degradation appearance and thermal performance;
- The samples tested during the wetting and drying processes exhibit behavioural differences, i.e., during the drying process it is possible to observe a homogeneous drying in all specimen thicknesses which in wetting process;
- There is an increase in the drying time constant for interfaced materials compared to monolithic materials. This result is very relevant because not taking the interface into account, i.e., considering them to be monolithic walls, can result in a significant error in the prediction of moisture transport in construction materials;
- The farther from the base the interface is located, the greater the drying time constant;
- The interface can significantly retarded moisture transport, i.e., the discontinuity of moisture content across the interface, which indicated that there was a difference in capillary pressure across the interface.

References

1. H.A. Iglesias, J. Chirife, *Handbook of Food Isotherms* (Academic Press, New York, USA, 1983)
2. D. Marinos-Kouris, Z.B. Maroulis, C.T. Kirannoudis, Computer simulation of industrial dryers. Drying Technol. **14**(5), 971–1010 (1996)
3. Z.B. Maroulis, G.D. Saravacos, *Food Process Design* (Marcel Dekker, New York, USA, 2003)
4. N. Mendes, P.C. Philippi, P.C., A method for predicting heat and moisture transfer through multilayer walls based on temperature and moisture content gradients. Int. J. Heat Mass Transfer **48**, 37–51 (2003)
5. Wilson, M.A., Hoff, W.D., Hall, C., Water movement in porous building materials—XIII absorption in two-layer composite. Build. Environ. **30**, 209–219 (1993)

6. M. Qin et al., Coupled heat and moisture transfer in multi-layer building materials. Constr. Build. Mater. **23**, 967–975 (2009)
7. L. Pel, Moisture transport in porous building materials. PhD Thesis, Technical University Eindhoven, The Netherlands (1993)
8. V.P. de Freitas, V. Abrantes, P. Crausse, Moisture migration in building walls—analysis of the interface phenomena. Build. Environ. **31**(2), 99–108 (1996)
9. L. Derdour, H. Desmorieux, J. Andrieu, A contribution to the characteristic drying curve concept: application to the drying of plaster. Drying Technol. **18**(1–2), 237–260 (2000)
10. T. Bednar, Approximation of liquid moisture transport coefficient of porous building materials by suction and drying experiments. Demands on determination of drying curve, in *Proceedings of 6th Symposium on Building Physics in the Nordic Countries*, Trondheim, Norway (2002), pp. 493–500
11. M. Karoglou, A. Moropoulou, Z.B. Maroulis, M.K. Krokida, Drying kinetics of some building materials. Drying Technol. **23**(1–2), 305–315 (2003)
12. F. Descamps, Continuum and discrete modelling of isothermal water and air transfer in porous media. Ph.D. Thesis, Catholic University of Leuven, Leuven, Belgium (1997)
13. M. Parti, Selection of mathematical models for drying grain in thin-layers. J. Agric. Eng. Res. **54**(4), 339–352 (1993)
14. C.L. Hii, C.L. Law, M. Cloke, Modelling of thin layer drying kinetics of cocoa beans during artificial and natural drying. J. Eng. Sci. Technol. **3**(1), 1–10 (2008)
15. C.L. Hii, C.L. Law, M. Cloke, Modeling using a new thin layer drying model and product quality of cocoa. J. Food Eng. **90**(2), 191–198 (2009)
16. A. Taheri-Garavand, S. Rafiee, A. Keyhani, Mathematical modelling of layer drying kinetics of tomato influence of air dryer conditions. Int. Trans. J. Eng. Manage. Appl. Sci. Technol. **2**, 147–160 (2011)
17. I.T. Toğrul, D. Pehlivan, Modelling of drying kinetics of single apricot. J. Food Eng. **58**(1), 23–32 (2003)
18. O. Corzo, N. Bracho, A. Pereira, A. Vásquez, Weibull distribution for modelling air drying of coroba slices, LWT—Food Sci. Technol. **41**(10), 2023–2028 (2008)
19. R. Baini, T.A.G. Langrish, Choosing an appropriate drying model for intermittent and continuous drying of bananas. J. Food Eng. **79**(1), 330–343 (2007)
20. A.S. Guimarães, J.M.P.Q. Delgado, A.C. Azevedo, V.P. de Freitas, Interface influence on moisture transport in buildings. Constr. Build. Mater. **162**, 480–488 (2018)
21. T. Colinart, P. Glouannec, Investigation of drying of building materials by single-sided NMR. Energy Procedia **78**, 1484–1489 (2015)
22. C. Nunes, L. Pel, J. Kunecký, Z. Slížková, The influence of the pore structure on the moisture transport in lime plaster-brick systems as studied by NMR. Constr. Build. Mater. **142**, 395–409 (2017)
23. C. Nunes, Z. Slížková, Hydrophobic lime based mortars with linseed oil: characterization and durability assessment. Cem. Concr. Res. **61**, 28–39 (2014)
24. R.S. Boynton, *Chemistry and Technology of Lime and Limestone*, 2nd edn. (Wiley, New York, USA, 1980)
25. N. Shahidzadeh-Bonn, A. Azouni, P. Coussot, Effect of wetting properties on the kinetics of drying of porous media. J. Phys.: Condens. Matter **19**(11), 112101 (2007)
26. X. Qiu, Moisture transport across interfaces between building materials. Ph.D. thesis, Concordia University, Canada, Montreal (2003)

Chapter 4
Conclusions

The development of this study contributes to the advancement in the knowledge of the behaviour of moisture transport in masonry walls with ceramic blocks. This study allows for evaluation and clarification of some aspects, namely: more frequent types of interfaces that masonry walls have; quantify and qualify the hygrothermal properties of constituent materials in masonry (ceramic bricks and mortars); the influence of these interfaces on the drying process, using two experimental methods (gravimetric method and gamma rays attenuation). The most significant conclusions drawn from the studies carried out are as follows:

- Regarding water transport phenomena in walls of different materials and different interfaces, in the first part of the study, it was possible to augment our knowledge concerning this area. After this step, materials that are mostly used in historical building walls were checked and the ceramic brick material with two densities was selected. The types of interfaces found in building walls are hydraulic contact (cement mortar and lime mortar), perfect contact and air space. These three were chosen to carry out the tests of this research.
- To quantify the amount of water that flows through the interface of building walls and to analyse the moisture transport in construction materials based on the use of the drying processes, tests were proposed and performed using two methods: gravitational method and gamma radiation. The first one is an easy-to-use method worldwide and the second involves a machine that was developed in the Laboratory of Building Physics (LFC). A moisture measuring device based on a non-destructive method of gamma rays attenuation facilitates measurements to deepen concepts in building physics related to the moisture transfer.
- Knowledge of building material properties is essential to predict heat, air and moisture transport in building elements and components, and for a correct characterization of the hygrothermal behaviour to predict pathologies. In this study, two important aspects were covered: (a) a laboratory characterization of material properties for two different ceramic brick blocks was carried out to determine their hygrothermal behaviour; (b) moreover, the material properties of mortars

J. M. P. Q. Delgado et al., *Drying Kinetics in Building Materials and Components*,
SpringerBriefs in Applied Sciences and Technology,
https://doi.org/10.1007/978-3-030-31860-4_4

(cement mortar and lime mortar) were studied to determine their hygrothermal and mechanical properties. Based on the experimental tests of this investigation, the following topics can be drawn for each study: (i) Ceramic brick blocks: The properties that were measured were bulk porosity, density, water vapour permeability, moisture storage function (moisture content), water absorption coefficient, moisture diffusivity and thermal conductivity; (ii) Mortars: The properties measured were not only the hygrothermal properties but also physical properties and mechanical properties, including a fine aggregate analyse. The determination of those properties was explained in depth. Those properties are commonly necessary as input for the hygrothermal simulation programmes to predict building materials and components' hygrothermal performance. With these results it is now possible to carry out several numerical studies to predict, for example, durability, waterproofing, degradation appearance and thermal performance of buildings or building components. If other authors want to study the hygrothermal behaviour of buildings or building components made by another materials, it will also useful to understand the way they calculate/determine the required hygrothermal properties or even physical properties and mechanical properties here explained.

• The results show that when the moisture reaches the interface there is a slowing of the humidifying process due to the interfaces hygric resistance. The samples with hydraulic contact interface (cement mortar) present lower absorption rate than the samples with lime mortar. The only situation where final absorption reaches the value of the monolithic test result is with an interface of hydraulic contact (lime mortar) and the highest value of hygric resistance was obtained with an interface of hydraulic contact (cement mortar). It was possible to observe the influence of air space between layers. If the layers of consolidated materials are separated by an air space interface, an initial constant absorption rate, instead a very slow absorption when the water reaches the interface was observed. The air space interfaces increase the coefficients of capillary significantly, as the distances from the contact with water increase.

• The study of building materials' drying kinetics is, therefore, very relevant for the construction industry to avoid damage or deterioration. The samples tested during the drying processes present a different behaviour, i.e., during the drying process it is possible to observe a homogeneous drying along the sample; not to include the interfaces in the building materials can result in significant prediction errors in the moisture transport simulation programs; There is an increase in the drying time constant for the materials with interface compared to the monolithic materials; as the further away from the base the interface is located, the greater the drying time constant; the interface could significantly retard the flow of moisture transport, i.e., the discontinuity of moisture content across the interface indicated that there was a difference in capillary pressure across the interface.

Printed in the United States
By Bookmasters